T0344260

Invitation to
Generalized
Empirical Method

In Philosophy and Science

It is necessary to distinguish in concrete relations between two components, namely, a primary relativity and other, secondary determinations.*

So it comes about that the extroverted subject visualizing extension and experiencing duration gives place to the subject orientated to the objective of the unrestricted desire to know and affirming beings differentiated by certain conjugate potencies, forms, and acts grounding certain laws and frequencies.†

---

* Bernard Lonergan, *Insight: A Study of Human Understanding*, vol. 3 in Frederick E. Crowe and Robert M. Doran, eds. *Collected Works of Bernard Lonergan* (Toronto: University of Toronto Press, 1992), 515. See "Metaphysics as Science," ch. 16, 512ff.
† Ibid, 537.

Invitation to
Generalized
Empirical Method

In Philosophy and Science

Terrance J Quinn

Middle Tennessee State University, USA

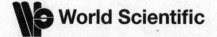

World Scientific

NEW JERSEY · LONDON · SINGAPORE · BEIJING · SHANGHAI · HONG KONG · TAIPEI · CHENNAI · TOKYO

Published by

World Scientific Publishing Co. Pte. Ltd.

5 Toh Tuck Link, Singapore 596224

USA office: 27 Warren Street, Suite 401-402, Hackensack, NJ 07601

UK office: 57 Shelton Street, Covent Garden, London WC2H 9HE

British Library Cataloguing-in-Publication Data
A catalogue record for this book is available from the British Library.

INVITATION TO GENERALIZED EMPIRICAL METHOD
In Philosophy and Science

ISBN 978-981-3208-43-8

Typeset by Stallion Press
Email: enquiries@stallionpress.com

Printed in Singapore

To my friends, family, colleagues and students.
And to scholars present and future
who share in the search and hope
for a balanced empirical method.

Preface

(I)t is obvious, that by our lights, there is no sharp line between philosophy of physics and physics itself.[1]

There is no clear boundary between philosophy of biology and theoretical biology.[2]

(S)ome philosophical work shades off into science; there is no sharp border between them.[3]

Perhaps what we mainly need is some subtle change in perspective – something that we all have missed.[4]

The quotations above are representative of a growing awareness in both the sciences and philosophy of science that, even though there are differences between philosophy of science and what is usually called science, it is also evident that science and philosophy of science are not separate. But, of course, that also begs the question. For, in order to solve contemporary problems, we will need precision, within up-to-date contexts, about how science and philosophy contribute to the whole enterprise. Penrose suggests that what is needed might be a subtle change in perspective. A purpose of the present book is to help bring out that a subtle change needed will be to a balanced empirical method that Bernard Lonergan called *generalized empirical method.*[5]

---

[1] Jeremy Butterfield and John Earman, eds. *Philosophy of Physics, Part A* (Amsterdam: Elsevier, 2007): vxiii.

[2] Brian Garvey, *Philosophy of Biology* (Stocksfield Hall: Acumen, 2007), xii.

[3] Peter Godfrey-Smith, *Philosophy of Biology* (Princeton: Princeton University Press, 2014), 1.

[4] Roger Penrose, *The Road to Reality, A Complete Guide to the Laws of the Universe* (New York: Alfred Knopf, 2005), 1045.

[5] Bernard Lonergan, *Insight: A Study of Human Understanding*, eds. Frederick E. Crowe and Robert M. Doran, vol. 3 in *The Collected Works of Bernard Lonergan* (Toronto: University of Toronto Press, 1992), 96–96. (First published: London: Longmans, Green and Co., 1957.) Henceforth, *CWL3*. Lonergan later provided a more precise definition in, Bernard Lonergan, *A Third Collection, Papers by Bernard J. F. Lonergan, S. J.* ed. by

P.1   Science and Philosophy of Sciences

Advances in philosophy of science have been made along various lines of thought. But, within philosophy of science as a whole, there is little sign of an emerging consensus. There is, instead, an ongoing multiplication of methods, interpretations and philosophic views. Since the early 20th century, philosophy of science has increasingly been looking to particular issues in the sciences (in, for example, physics, chemistry, biological sciences, mathematical sciences, mathematical logic, computer sciences, engineering sciences and medical sciences). Despite that focus, however, the tradition does not require that contributions to the literature be verifiable in scientific practice.[6]

This is causing problems in contemporary philosophy of science. But, is it a problem for the sciences? Some suggest not, that philosophical issues needn't be of concern to, or bear on scientific development; that philosophy of science has its own agenda; and that science can get on with its own business without needing input from philosophy of science. There are, however, evident problems with such views.

One way to begin to see this is to note that, whether adverted to or not, at every stage of development, we each have our own more, or less, nuanced openness (or not) and willingness (or not), to kinds of problem

---

Frederick E. Crowe, S. J. (New York: Paulist Press, 1985), 141. Henceforth, *A Third Collection*. See Section 2, for description of the new method.

[6] "The scientific effort to understand is blocked by a pretense that one understands already, and indeed in the deep, metaphysical fashion" (*CWL3*, 528–529). That approach to philosophy and philosophy of science has some roots in the work of Bernard Bovier de Fontenelle (1657–1757), who promoted an 'haute vulgarization.' Fontenelle wrote philosophy for the "worldly salons, …, (whom) he regarded … as his essential audience" (Steven F Rendall, "Fontenelle and his Public," *Modern Language Notes* 86 (4) (1971): 496–508.) In his own words (translated): "To penetrate into things either obscure in themselves, or but darkly expressed, requires deep Meditation, and an earnest application of the Mind; but here nothing more is requisite than to Read, and to imprint an idea of what is read, in the Fancy, which will certainly be clear enough (Bernard le Bovier, M. de Fontenelle, *Conversations on the Plurality of Worlds*, trans. William Gardiner, Esq. (London: A. Bettesworth, 1715), page A4. See also, *Classic eBook Collection*, Open Library, https://openlibrary.org).

and types of question; to views of one kind and another; to kinds of criteria needed (or not); to what results might mean (or not); to what is real (or not); to reasonable strategies and choices (or not); and, generally, to what might contribute to progress in science and society (and what probably will not).

At the same time, as both science and philosophy of science continue to develop, the complexity and subtlety of issues also increases. No doubt, we can describe apparent differences between the sciences and philosophy of science. But, it is also now generally recognized that there are no separations as such — neither between sciences; nor between sciences and philosophy of science; nor, likewise, between developing, as well as emerging (pure and applied) sciences and technologies. In fact, it is becoming increasingly obvious that, in ways that have yet to be identified,[7] scientific progress is part of a geo-historical omnidisciplinary collaborative enterprise.[8]

However, present traditions of science and philosophy of science do not educate one to advert to one's own experience, to seek to identify how what we do in one area might relate to, or complement how we get results in other areas. Mainly, the focus is object-oriented. For the individual, there is a self-screening. At the community level, this allows for, among other things, more or less endless debate. And, with issues increasingly complex, a lack of self-knowledge also contributes to confusion in front-line field work and interdisciplinary collaboration.

That there are difficulties is especially obvious in sciences such as biochemistry, embryology and developmental biology, neuroscience and animal psychology, ecology and the environmental sciences — that is, where sciences are finding complex flexible "layerings" of properties (physical; chemical, and so on) and vast aggregates of events and occurrences. What about physics? Physics is well-known for its remarkable progress over the last few centuries. But, it is not exempt from

---

[7] At this time, it is generally acknowledged that "the discussion of interdisciplinarity in the philosophy of science is still in its infancy" (Till Grüne-Yanoff, "Introduction: Interdisciplinary model exchanges," *Studies in History and Philosophy of Science*, vol. 48 (2014): 59).

[8] See Epilogue.

the challenge. It too struggles with similar difficulties, with mutually incompatible clusterings of views within and about its mission;[9] and a lack of consensus about the significance of physics in, for example, physical-chemistry; biophysics; emergence and the cosmos.[10]

P.2   *Insight*[11]: A Potentially Helpful Context

Readers already familiar with the work of Bernard Lonergan will not be surprised that I suggest that we take help from the book *Insight*.[12] In his book, Lonergan briefly described a needed development in empirical method, what he called *generalized empirical method*:

> We have followed the common view that empirical science is concerned with sensibly verifiable laws and expectations. If it is true that essentially the same method could be applied to the data of consciousness, then respect for ordinary

---

[9] Subtly implicit in contemporary theories are various confusions, some of which have roots in the work of 20th century founders such as Schrödinger and others. Helpful here is, Philip McShane, "Elevating *Insight*: Space-Time as Paradigm Problem," *Method: Journal of Lonergan Studies*, vol. 19, no. 2 (2001): 203–229. Other confusions, more visible in that at present there is open debate about method, regard the plausibility (or not) of string theory. (The string theory story began with a formula discovered in 1968 by an Italian physicist, Gabriele Veneziano. Veneziano "saw an interesting pattern in the data. He described the pattern by writing down a formula that described the probabilities for two particles to scatter at different angles. Veneziano's formula fit some of the data remarkably well" (Lee Smolin, *The Trouble with Physics, The Rise of String Theory, The Fall of a Science, and What Comes Next* (Boston: Houghton Mifflin, 2006) 103). Advocates of string theory argue that philosophical argument together with mathematics and aesthetics is sufficient to assert that string theory explains physical geometry. This is in contrast to a centuries-long community effort that gave rise to the present Standard Model. In other words, besides aspirations of string theory, there are traditional empirical methods wherein experimental physics looks for anomalies in data and theoreticians seek to explain those anomalies.

[10] The current literature is large. But, see, for example, the spread of views found in John D. Barrow, Paul C. W. Davies and Charles L. Harper, Jr, eds., *Science and Ultimate Reality. Quantum Theory, Cosmology, and Complexity*. Cambridge: Cambridge University Press, 2004.

[11] *CWL3*.

[12] *CWL3*.

usage would require that a method which only in its essentials is the same be named a generalized empirical method.[13]

In the years following *Insight*,[14] Lonergan continued to investigate methods in the sciences, philosophy and theology.[15] In 1974, he provided a more precise definition:

Generalized empirical method operates on a combination of both the data of sense and the data of consciousness: it does not treat of objects without taking into account the corresponding operations of the subject; it does not treat of the subject's operations without taking into account the corresponding objects.[16]

One of the advantages of such a method will be the possibility of a new and developing "control of meaning."[17] But, if we are to take help from *Insight*, we need to avoid the trap of rich description. The book *Insight* is, to some extent, an impossible book. Its emergence in history is an anomaly. The chapters climb steeply to a heuristics of the empirical sciences, philosophy of science, and general philosophy. From the beginning, the first chapters of *Insight* challenge the reader to advert to their experience in examples from across more than two millennia of mathematics, physics, and other sciences.[18] Perhaps, though, chapters 6 and 7, on common sense, are more accessible? But,

it would be impossible for common sense to grasp what precisely common sense happens to illustrate.[19]

At the end of Chapter 1 of *Insight*, Lonergan himself indicates something of the challenge of reading the book:

---

[13] *CWL3*, 96.

[14] *Insight* was first published in 1957. See note 5.

[15] See, for example, Parts Two and Three of, Bernard Lonergan, *A Third Collection*.

[16] Bernard Lonergan, *A Third Collection*.

[17] *CWL3*, 530, in sec. 16.3.4, "The Significance of Metaphysical Equivalence," 530–533.

[18] (T)he precise nature of the act of understanding is to be seen most clearly in mathematical examples; the dynamic context in which understanding occurs can be studied to best advantage in an investigation of scientific methods; the disturbance of that dynamic context by alien concerns is thrust upon one's attention by the manner in which various measures of common nonsense blend with common sense" (*CWL3*, 4).

[19] *CWL3*, 15.

(I)nstead of the story of Archimedes the reader will profitably substitute some less resounding yet more helpful experience of his own. Instead of the definition of the circle he can take any other intelligently performed act of defining and ask why the performance is, not safe, not accurate, not the accepted terminology, but a creative stroke of insight. Instead of the transition from elementary arithmetic to elementary algebra one may review the process from Euclidean geometry to Riemannian geometry. Instead of asking why surds are surds, one can ask why transcendental numbers are transcendental. Similarly, one can ask whether the principle of inertia implies that Newton's laws are invariant under inertial transformations, what inspired Lorenz to suppose that electromagnetic equations should be invariant under inertial transformations, whether an inverse insight accounts for the basic postulate of special relativity, whether the differences of particular places or particular times are the same aspect of the empirical residue as the differences of completely similar hydrogen atoms. For, just as in any subject one comes to master the essentials by varying the incidentals, so one reaches familiarity with the notion of insight by modifying the illustrations and discovering for oneself and in one's own terms the point that another attempts to put in terms that he happens to think will convey the idea to a probably nonexistent average reader.[20]

That paragraph, sitting quietly at the end of Chapter 1 of *Insight*, invites a new way of reading and writing that gradually will transform the Academy. How, then, can we take help from the book *Insight*? I suggest that we take the advice of the author of *Insight*.

The book *Insight* is descriptive of an "arduous and exploratory journey"[21] taken by Lonergan, but not yet by us. The book is doctrinally compact. It is a nuanced and abbreviated telling of a genius-lone-climb to a control of meaning remote to present-day scholarship. The chapters of *Insight* speak of progress and decline, growth and development. But, the book asks that the reader climb, through a series of exercises. Reading the book without doing the exercises is not reading the book the way the author intended. But, taking the pointing of the last paragraph of Chapter 1 of *Insight* to heart, development in generalized empirical method will be in our own terms[22] in, among other things, our experience in the developing sciences.

---

[20] *CWL3*, 56.
[21] *CWL3*, 12.
[22] See note 20.

P.3 Near Future Needs; and Remote Future Contexts

In order to promote emergence of the balanced empirical method, we will need diverse series of works that creatively enter into sometimes elementary, and sometimes more advanced examples, problems and topics within the developing sciences, technologies, philosophy of science technology and education.[23] Initially, this mainly will be learning and exploring potentialities of the new method. For a time, needed works mainly will be transitional, to help seed and cultivate emergence of balanced empirical method.

Eventually, though, as the new method becomes normalized, there will be cultural advances that cannot yet be imagined. In particular, pedagogies from Kindergarten to graduate school will be transformed to raise students to be present to subtleties of their own experiencing; wondering; understanding; judging; deliberating; and deciding.

P.4 Present Context and Purpose of the Book

It is about 60 years since the first edition of *Insight*.[24] In that time, some scholars have, in various ways, discussed and explored the possibility of the new method.[25] However, generalized empirical method will be a

---

[23]A major contribution already available are the writings of Philip McShane. See, http://www.philipmcshane.org/. However, what is needed will be contributions of many authors and teachers, multi-generational, and in all areas.

[24]Bernard Lonergan, *Insight: A Study of Human Understanding*, first pub. in London: Longmans, Green & Co., 1957.

[25]A sampling from the literature includes: Michael Shute, *Lonergan's Discovery of the Science of Economics*. Toronto: University of Toronto Press, 2010. Frank E. Budenholzer, "Emergence, Probability and Reductionism," *Zygon*, vol. 39, no. 2 (June, 2004): 339–355; Mark D. Morelli, "Obstacles to the Implementation of Lonergan's Solution to the Contemporary Crisis of Meaning," in John J. Liptay Jr. and David S. Liptay, *The Importance of Insight, Essays in Honour of Michael Vertin* (Toronto: University of Toronto Press, 2007), 22–48; Patrick Byrne, "Statistics as Science: Lonergan, McShane, and Popper," *Journal of Macrodynamic Analysis*, vol. 3 (August, 2003): 55–75; Donna Teevan, "Albert Einstein and Bernard Lonergan on Empirical Method," *Zygon* 37 (Dec. 2002): 873–90; Ivo Coelho. *Hermeneutics and Method: The 'Universal Viewpoint' in Bernard Lonergan*. Toronto: University of Toronto Press, 2001; Hugo A. Meynell, *An Introduction to the Philosophy of Bernard Lonergan*, 2nd ed. Toronto: University of Toronto Press, 1991

community achievement. And, so far, it is not yet operative. A main purpose of this book is to help further reveal the need and possibility of the new method. At the same time, the book is to help toward bringing out the need of series of books such as those mentioned in Section 3.

P.5 Strategy of the Book

The book, then, does not illustrate generalized empirical method, but is an invitation to a development in method. In that sense, the book is no more about Lonergan than an invitation to chemistry is about Mendeleev. The book invites readers to advert to various strategically ordered examples in the sciences and philosophy of science. These are provided not to engage in, or blend the present aim with prior forms of philosophical debate. Examples are provided as points of entry. Some are discussed in more detail than others. While samples of writings representing different philosophical views are given, there are no comparisons of the type 'Heisenberg and Dirac' or 'Lonergan and Kuhn.' Likewise, there are no alleged "dialogues" between authors A and B — such *dialogue*, of course, being only metaphor. It is not that interpretation and comparison of the works of authors are not essential parts of ongoing scholarship. But, new

---

(1st ed., London: MacMillan Press, 1976); Kenneth Melchin, *History, Ethics, and Emergent Probability: Ethics, Society, and History in the Work of Bernard Lonergan.* Lanham, MD: University Press of America, 1987; William J. Danaher, "Lonergan and the Philosophy of Science," *Australian Lonergan Workshop* (1985), 31–46; Philip McShane, "The Importance of Rescuing *Insight*," in Liptay and Liptay (2007), 199–226; Philip McShane, "Obstacles to Metaphysical Control of Meaning," *Method: Journal of Lonergan Studies*, vol. 23, no. 2 (2005): 187–195; Philip McShane, "Elevating Insight: Space-Time as Paradigm Problem," *Method: Journal of Lonergan Studies*, vol. 19, no. 2 (2001): 203–229; Philip McShane, "Scientific Methods and the Investigation of Ultimate Meanings," *Journal of Ultimate Reality and Meaning*, vol. 11 (1988): 142–44; Philip McShane, *Randomness, Statistics and Emergence.* Notre Dame, IN: University of Notre Dame Press, 1970; Philip McShane, "Insight and the Strategy of Biology," *Continuum* 2 (1964): 374–88; and "The Foundations of Mathematics," *Modern Schoolman*, vol. 40 (1963): 373–87. See also the many works of Philip McShane, available at http://www.philipmcshane.org/, the collection of which constitutes a major contribution to seeding series of works needed, mentioned in Section 3.

kinds of interpretation and comparison will emerge later within the new method.[26]

As Melchin notes of Vertin's work, the reader of the present book is invited to

> cast in personal terms, ..., personal observations ... on operations ... for the evidential base for ... arguments and analyses.[27]

The context and discussion of the present book is, though, modern science and modern philosophy of science. Through preliminary self-attention, one can begin to (self-) notice problems in contemporary philosophic methods. But, positively, examples also are to help toward beginnings in realizing the possibility and advantages of a balanced empirical method for science and philosophy of science.

The book climbs in difficulty. This is not in order to reach conclusions. The climb is to help bring out some of the many dimensions of the problem. And, beginnings toward the new method will be made by exploring examples.

---

[26] Lonergan looks to future methods of interpretation: "The totality of documents cannot be interpreted scientifically by a single interpreter or even by a single generations of interpreters. There must be a division of labor, and the labor must be cumulative. Accordingly, the fundamental need is for reliable principles of criticism that will select what is satisfactory and will correct what is unsatisfactory in any contributions that are made" (*CWL3*, 610). Note that comparison presupposes interpretation. Does one compare Einstein's work on General Relativity with Eddington's, without being at home in General Relativity? Or, there is the problem of attempting to compare Lonergan's views with Kuhn's, say. Does not such an effort call for a control of meaning comparable to what Lonergan attained, one of the authors in the comparison? The high challenge of explanatory interpretation is pointed to in Lonergan, *Insight*, sec. 17.3. See "canon of explanation," 609–610. See also "Interpretation," ch. 7 in, Bernard Lonergan, *Method in Theology*. London: Darton, Longman and Todd, 1972/73/75. Both comparison and interpretation of works of authors will become effective within a future functional division of labor. See Epilogue.

[27] Ken Melchin, "Democracy: Sublation, and the Scale of Values," in Liptay and Liptay, 185.

P.6 Intended Audience

The book is for readers of various academic backgrounds. For instance, it will be appropriate for those who already have made some progress in a "basic position,"[28] but who wish to follow up with examples from the sciences. For those scholars already familiar with some of the difficulties and advantages of self-attention, the book extends the invitation to "supply (one's) philosophy with instances."[29] The book may be of interest to mid-career or late-career scholars in Lonergan Studies who, on reading the book, might then encourage their students or their students' students to begin early in discovering the joy of scientific development, as well as the fundamental relevance of scientific development to contemporary philosophy.[30] The book also is intended for readers in the sciences who are interested in foundational issues. I expect that some readers from the sciences may find the approach unusual, for, there is

the climate of the times, which in many ways discourages serious self-attention.[31]

Readers trained in contemporary analytic philosophy of science may have

---

[28] "It will be a basic position (1) if the real is the concrete universe of being and not a subdivision of the 'already out there now'; (2) if the subject becomes known when it affirms itself intelligently and reasonably and so is not yet known in some prior 'existential' state; and (3) if objectivity is conceived as a consequence of intelligent inquiry and critical reflection, and not as a property of a vital anticipation, extroversion, and satisfaction" (*CWL3*, 413).

[29] *CWL3*, 455.

[30] "The contribution of science and of scientific method to philosophy lies in a unique ability to supply philosophy with instances of the heuristic structures which a metaphysics integrates into a single view of the concrete universe" (*CWL3*, 455). "(M)etaphysics regards proportionate being as explained" (*CWL3*, 515).

[31] Mark Morelli, "Obstacles to Implementation of Lonergan's Solution to the Contemporary Crisis of Meaning," in Liptay and Liptay, 45. Regarding "the times," see Bernard Lonergan, *Method in Theology*, "Stages of Meaning," sec. 3.10, 85–99. The present times are a second stage of meaning, a stage of rapid development, increasing confusion, partial progress and also cumulative decline. There is the possibility, however, of humanity reaching a third stage of meaning (*Method in Theology*, 85ff). Key to transitioning to a third stage of meaning will be the gradual emergence and survival of functional collaboration. See Epilogue.

similar difficulties with the book. Contemporary philosophy tends to emphasize terminologies, conceptual constructs and logical analysis. But, part of what is promoted in this book is the emergence of a balanced empirical method which will include attending to one's acts and operations.

Acknowledgments

Much of Chapter 2 first appeared in "Fledgling Functional Foundations for the Biology of the Adult Pigeon," *Revista de Filosofía,* Año 45, no. 135, Universidad Iberoamericana, México (2013): 123–152.

Thanks to the World Scientific Publishing Acquisitions Editor, Chad Hollingsworth (Hackensack, NJ); and the Desk Editor, Low Lerh Feng (Singapore). Chad helped make the book proposal and review process go smoothly. Lerh Feng was a great help to me in the production of the final manuscript. Their assistance was much appreciated.

I would like to thank the anonymous referees for their constructive comments; as well as Michael Shute, for helpful suggestions.

Special thanks to Philip McShane, for his support and encouragement through the years.

Thanks, finally, to Sreten Marilovic and Sue Mumford for their ongoing friendship and support behind the scenes, in particular, during my break from teaching when I started working on this book in Toronto in the autumn of 2012.

Contents

Introduction

I.1 Generalized Empirical Method

Lonergan discovered the possibility of a *generalized empirical method*[1] that "envisages all data."[2] As noted by Lawrence, eventually it will be known simply as adequate "Empirical Method."[3] And yet: Is generalized empirical method needed, and feasible? Might there be advantages in such a method? Will it solve problems?

Attempts to implement the method in the developing sciences reveals that it will be challenging, subtle and new. What will the new method look like in, for example, investigations into ancient geometry and physics, let alone late 19th century mathematics and physics, basic chemistry, or more recent quantum field theories and quantum chemistry? What of biology, where in embryology, for example, one investigates the emerging neurodynamics of a developing avian embryo? Or, again, there are ecology and environmental science, areas of inquiry that investigate vast aggregates of events and occurrences in ecosystems.

Here, we might recall Aristotle's observation:

For the things we have to learn before we can do them, we learn by doing them.[4]

---

[1] Bernard Lonergan, *Insight: A Study of Human Understanding, Collected Works of Bernard Lonergan*, vol. 3., eds. Frederick E. Crowe and Robert M. Doran, Toronto: University of Toronto Press, 1992 (1st ed.: London: Longman, Green & Co., 1957). Henceforth *CWL3*. Bernard Lonergan, "Second Lecture," ch. 9 in, *A Third Collection*, Papers by Bernard J. F. Lonergan, ed. Frederick E. Crowe, S. J. (New York: Paulist Press, 1985), 141. Henceforth, *A Third Collection*. See Preface, Section P.2, for both descriptions of *generalized empirical method*.

[2] *A Third Collection*, 140.

[3] Fred Lawrence, "The Ethics of Authenticity and the Human Good," in John J. Laptay Jr. and David S. Laptay, eds., *The Importance of Insight: Essays in Honor of Michael Vertin* (Toronto: University of Toronto Press, 2007), 131.

[4] Aristotle, *The Nicomachean Ethics*. tr. by W. D. Ross (Blacksburg, VA: Virginia Tech, 2001), Book II, par. 1.

For, it is, in fact, too soon to be able answer questions such as those posed above. We need, first, to enter the developing modern sciences, and there make preliminary efforts with particular cases and, that way, learn by doing.

I.2 Preliminary Context of the Book

In the 20th century, philosophy of physics emerged as a major zone of scholarship.[5] In recent decades, scholars in philosophy of physics[6] have been grappling with sophistications of multi-variable probability theories, thermodynamics, quantum theories, special and general relativity, gauge theories, and more recently, superstring theories. And, advances are being made along various lines of thought.

A basic problem, however, is that there is little sign of an emerging consensus. There are, instead, ongoing debates and expanding ranges of views - some of which partially overlap, while others are (implicitly or explicitly) mutually incompatible.[7] A problem here is not that there are so *many* views, but that philosophy of physics seems to have no way to effectively promote progress toward shared heuristics. Important insights regularly are reached by leading scholars. But, rebuttals tend to be admissible so long they are logical and have coherent terminologies. This allows for more or less endless debate wherein views need have little or no contact with scientific practice, data, or actual objects of scientific inquiry.

---

[5] For a short description of philosophy of physics in the 20th century, see Introduction, in, Jeremy Butterfield and John Earman, eds., *Philosophy of Physics (Handbook of the Philosophy of Science)*, Part A, of 2 vol. set, 1st ed. Amsterdam: Elsevier, 2007.

[6] See, for example, Christian Wüthrich, webpages "Philosophers of Physics," and "Library," *Taking up Spacetime*, https://takingupspacetime.wordpress.com/philosophers-of-physics-the-websites/.

[7] For example, see, John D. Barrow, Paul C. W. Davies and Charles L. Harper, Jr, eds., *Science and Ultimate Reality. Quantum Theory, Cosmology, and Complexity*. Cambridge: Cambridge University Press, 2004; and Martin Curd, J. A. Cover and Christopher Pincock, eds., *Philosophy of Science. The Central Issues*, 2nd ed. New York: W. W. Norton and Co., 2013.

Writing about present-day challenges in physics, John Earman refers to

> clouds on the horizon that may prove as great a threat to the continued success of twentieth century physics, as were the anomalies confronting classical physics at the end of the nineteenth century.[8]

But, in contemporary philosophy of physics there is an ethos that allows the scholar to back off from taking a stand on basic questions:

> As philosophers we are generalists: so we naturally find all the various foundational issues mentioned above worrisome. But, being generalists, we will of course duck out of trying to say which are closest to the solution, or which are most likely to repay being addressed. In any case, such judgments are hard to adjudicate, since intellectual temperament, and the happenstance of what one knows about or is interested in, play a large part in forming them.[9]

While contemporary philosophy of physics struggles without a normative heuristics grounded in scientific practice, it is also true that contemporary experimental and theoretical physics, for their part, do not always pay much attention to the results of philosophic reflection. An extreme case is the view of expressed by cosmologist Lawrence Krauss:

> (T)he worst part of philosophy is the philosophy of science; the only people, as far as I can tell, that read work by philosophers of science are other philosophers of science. It has no impact on physics whatsoever, and I doubt other philosophers read it, because it's fairly technical. And so it's really hard to understand what justifies it. ... (S)cience progresses and philosophy doesn't.[10]

Krauss, however, distinguishes

---

[8] Butterfield and Earman, Part A, xx.

[9] Butterfield and Earman, Part A, xx-xxi.

[10] Victor J. Stenger, "Physicists are Philosophers, Too," *Scientific American* (May 8, 2015): par. 1. http://www.scientificamerican.com/article/physicists-are-philosophers-too/. Krauss may have changed his views. The quotations, though, remain convenient, for they are representative of a common bias.

questions that are answerable and those that are not,[11]

and asserts that answerable questions mostly fall into the

domain of empirical knowledge, aka science.[12]

Besides the obvious performance-contradiction of Krauss' claim, there are, in fact, complex mesh-works of subtle views regarding the importance (or not) and relevance (or not) of philosophy of physics to physics. Sometimes views are implicit, influencing work in rather hidden ways. But, some physicists have made the effort to explain and justify anti-philosophy views. So, Stenger, too, then draws attention to performance-contradiction. As in Krauss' view, there is what anti-philosophy views say. But, what do scholars do in order to reach and defend such views?

> (P)rominent critics of philosophy … think very deeply about the source of human knowledge. That is, they are all epistemologists.[13]

Of course, not many physicists attempt to work out explicit views of knowledge, objectivity and reality. It may never be necessary that a majority of physicists do so. But, as alluded to in the previous paragraph, whether one adverts to what one does or not, we each have our views; and, at any time, we each have our openness (or not), to types of question that we deem worth asking (or not). All of this shapes ongoing inquiries, discoveries, developments, plans and strategies. So, while not in the sense necessarily of spelling out explicit views, at least in the sense of what

---

[11] Julian Baggini and Lawrence Krauss, "Philosophy v science: which can answer the big questions of life?", *The Guardian* (September 8, 2012, modified May 21, 2014): http://www.theguardian.com/science/2012/sep/09/science-philosophy-debate-julian-baggini-lawrence-krauss. See also, Ross Anderson, "Has Physics Made Philosophy and Religion Obsolete?" (Interview with Lawrence Krauss), *The Atlantic* (April 23, 2013): http://www.theatlantic.com/technology/archive/2012/04/has-physics-made-philosophy-and-religion-obsolete/256203/. Krauss may change his views of philosophy of science. The quotations, though, are convenient, for they represent a common and surviving bias.

[12] Julian Baggini and Lawrence Krauss, "Philosophy v science."

[13] Stenger, last paragraph.

physicists do, the claim of the title of Stenger's article (self-) evidently holds: "Physicists are philosophers, Too."[14]

Other examples reveal further aspects of progress in physics not adverted to by anti-philosophy views. Krauss and others speak of "empirical knowledge." But, what is *empirical knowledge*? Even in a so-called "hard-science" like physics, it is not only experimental data that contributes to progress. Are not Einstein's writings (old and more recently discovered[15]) empirical? In the ongoing effort to make new progress, Einstein's works are studied partly in the hopes of improving understanding of what Einstein meant, and what he didn't mean. The emergence of gauge theories, and eventually the Standard Model in particle physics, was a slow and tortuous journey.[16] Theoreticians such as Weyl, Einstein, Klein, Pauli, Feynman and others appealed to up-to-date experimental results. But, they also read, struggled with, argued with, and learned from each other's writings - the meanings and significance of which were not, and still are not, immediate. Obviously, then, progress in physics depends on ongoing efforts to obtain and interpret not just particle tracks on computer screens, but screen-tracks, print-tracks and write-tracks that are expressions of physicists working in diverse areas of the field. Whether adverted to or not, in every area of physics, expressions of physicists are empirical; and interpreting them is part of progress. And no area in modern physics can escape the challenges of progress in hermeneutics, epistemology and ontology.

Regarding struggles in physics of the last thirty years, Rovelli comments:

> Of course there are ideas. These ideas might turn out to be right. Loop quantum gravity might turn out to be right, or not. String theory might turn out to be right, or not. But we don't know, and for the moment, nature has not said yes in any sense. ... I suspect that this might be in part because of the wrong ideas we have about

---

[14] Stenger.

[15] *Einstein Papers Project, The Collected Papers of Albert Einstein*, http://www.einstein. caltech.edu/.

[16] Lochlainn O'Raifeartaigh and Norbert Straumann, "Gauge theory: Historical origins and some modern developments," *Reviews of Modern Physics*, vol. 72, issue 1 (2000): http://dx.doi.org/10.1103/RevModPhys.72.1; and Lochlainn O'Raifeartaigh, *The Dawning of Gauge Theory*. Princeton, NJ: Princeton University Press, 1997.

science, and because methodologically we are doing something wrong, at least in theoretical physics, and perhaps also in other sciences.[17]

For other sciences, we can, for example, look to biology. There too, we find numerous fundamentally opposed views, openly at odds with each other, clashing within and across areas. There are, for example, "dueling discourses in interdisciplinary biology."[18] As in physics, this too is not merely philosophical. For here too, ranges of different views guide experimental work, shape questions and interpretations, and are part of pedagogies that lead to the next generation of investigators and educators with similarly opposed views.

In the biological sciences, problems are especially evident in the groupings of views generally known as "systems biology."[19] Systems

---

[17]John Brockman, ed., "Science is Not About Certainty: Philosophy of Physics, A Conversation with Caro Rovelli," with Introduction by Lee Smolin. *Edge* (May, 2012): http://edge.org/conversation/carlo_rovelli-science-is-not-about-certainty-a-philosophy-of-physics.

[18]Jane Calvert and Joan H. Fujimura, "Calculating life? Dueling discourses in interdisciplinary systems biology," *Studies in History and Philosophy of Biological and Biomedical Sciences*, 42 (2011): 155-163.

[19] Institute for Systems Biology, http://www.systemsbiology.org/. General Systems Theory goes back to Bertanlanffy (1901-1972) and some of his contemporaries. They tried to account for organisms and societies in terms of axiomatic systems: "In rigorous development, general systems theory would be of an axiomatic nature; that is, from the notion of 'system' and a suitable set of axioms propositions expressing system properties and principles would be deduced" (Ludwig von Bertalanffy, *General System Theory, Foundations, Development, Applications*, rev. ed. New York: George Braziller, 1968), 55. See also Ervin Laszlo, *The Systems View of the World, The Natural Philosophy of the New Developments in the Sciences*. New York: George Braziller, 1972; Howard H. Pattee, ed., *Hierarchy Theory, The Challenge of Complex Systems*. New York: George Braziller, 1973; John W. Sutherland, *A General Systems Philosophy for the Social and Behavioral Sciences*. New York: George Braziller, 1973; et al. For a popular account of systems theory today, see Lars Skyttner, *General Systems Theory: Problems, Perspectives, Practice*. Hackensack, NJ: World Scientific Press, 2005. The systems theory viewpoint has been gaining influence in the scientific community. For a more technical presentation of contemporary results, see, for example, A. J. Marian Walhout, Marc Vidal and Job Dekker, eds., *Handbook of Systems Biology, Concepts and Insights*. Amsterdam: Academic Press, 2013. Nicolas Rashevsky (1899-1972) was a parent of systems theory in mathematical

biology seeks to understand organisms in terms of "systems," "levels of systems" and "biological information." One of the potential advantages of the approach is that it embraces interdisciplinarity from the start. This has led to developments in understanding complex combinations of layerings of time-dependent probability distributions in organisms and ecosystems, and is contributing to ongoing progress in the medical sciences and environmental sciences.

But, there are foundational problems in premises, conclusions and goals of systems biology. A main focus in contemporary systems biology is numerical computation and computer simulation of: information, control, levels, systems and subsystems. However, as will be revealed in the chapters below, these conceptual constructs are not verified in actual organisms (even in one-celled organisms, let alone multi-cellular plants, animals, and ecosystems). In the meantime, some of the main hypotheses of systems biology have contributed to the rise of horrific models wherein organisms allegedly are accounted for in terms of codes, molecular information tapes, genomic computer programs and other fictions.[20]

Commenting on the diversity of competing views in the academy as a whole, Rovelli offers hope:

> Somehow cultures reach, enlarge. I'm throwing down an open door if I say it here, but restricting our vision of reality today on just the core content of science or the core content of humanities is just being blind to the complexity of reality that we can grasp from a number of points of view, which talk to one another enormously, and which I believe can teach one another enormously.[21]

---

biology. Among other things, he studied the physics of cell membrane motion. See Section 2.2.

[20] For example: "There is no concept more intrinsic to systems developmental biology than that of process control by a preformed genomic program, in exactly the same sense as the term 'program' is used for the code directing a complex stepwise computational program" (Isabelle S. Peter and Eric H. Davidson, "Transcriptional Network Logic: The Systems Biology of Development," ch. 11 in, A. J. Marian Walhout, Marc Vidal and Job Dekker, eds., *Handbook of Systems Biology, Concepts and Insights* (Amsterdam: Academic Press, 2013), 213). See Chapter 2 for discussion.

[21] See note 17, last paragraph.

How, though, might we actually change how we are doing things so that methodologically we are doing what is right?[22] How, actually, might it be possible to "grasp from a number of points of view, which talk to one another enormously, and which ... can teach one another enormously"[23]? Whatever we do, we should not undermine progress that science and philosophy of science already are making. Yet, whatever change is possible, it also will need to help us get beyond present-day confusions and conflicts within and between sciences and philosophy of science.

Part of what is needed is a transitioning to a balanced empirical method, that is, *generalized empirical method.*[24] Generalized empirical method is not yet operative in the sciences and philosophy of science. No doubt, emergence of the balanced method will be difficult, and may well require a prolonged transition period. However, once normalized, generalized empirical method will be key not only to resolving old problems, but also to opening up new growth trajectories in science and philosophy of science. Among other things, the new method will be generative of new standards of competence and a *control of meaning*[25] that as much as possible will be "at the level of the times."[26]

I.3 Structuring of the Book

Some topics mentioned in earlier chapters ultimately depend on future progress discussed in later chapters. But, all topics are introduced for future follow-up in the academy. The book is not a treatise reaching conclusions but, as mentioned in the Preface, is an invitation to a development in empirical method.

Readers may be of various academic backgrounds. While the chapters climb in difficulty, I have attempted to make the main body of the text generally accessible. Footnotes include pointings to more advanced issues,

---

[22] See note 17.

[23] See note 17, last paragraph.

[24] See note 1.

[25] *CWL3*, 530, in sec. 16.3.4, "The Significance of Metaphysical Equivalence," 530-533.

[26] Frederick E. Crowe, *Lonergan* (Collegeville, MN: A Michael Glazier Book, 1992), chs. 3 and 5.

as well as references to up-to-date literature. All of this added to the complexity of the book structure. But, in the spirit of the main purpose of the book, it seemed important to provide references for the larger context.[27]

Regularly, there are quotations from the book *Insight*. This is not to invoke *Insight* as an authority on, say, correct terminology, or as a way to settle debates. However, as mentioned in the Preface, the book *Insight* can be helpful in providing precise (doctrinally compact) description that points to future developments. I sometimes write *self-attention*; and also sometimes refer to the possibility of being *luminous* in one's work. The term *self-attention* does not appear in the book *Insight*. However, generalized empirical method will depend on appealing to all data, all experience. That broadened focus of attention includes what can be called *self-attention*.[28] What begins to emerge is the need and possibility of a developing *control of meaning*[29] in the sciences and philosophy of science. It is convenient to have a name for this: *self-luminous*, or (within a later adequate Empirical Method[30]) *luminous*. While the possibility and need of being luminous begins to emerge, the normalized achievement will be something for future sciences and philosophy of science.

It will be noticed that much of the discussion draws on topics from biology. This is not to give priority to biology, or to suggest that biology is more accessible than, say, physics or chemistry. The emphasis on biology partly is because biology is an inviting context. But, as the book brings out, biology also provides a convenient context from which to enter many aspects of what, in fact, is an omnidisciplinary challenge.

Choices had to be made about which topics to include. As indicated in the Preface, the book reveals the need of suites of transitional works to be written by teams of authors. For instance, graduated series of pedagogical works will be needed on our "notion of a thing."[31] Note that the present book also moves past challenging foundational problems in chemistry. For

---

[27] All of this can be navigated according to one's interests and background.
[28] See, for example, *CWL3*, 3, 4, 5, 13, 14, 95-96, 422-423.
[29] See note 25.
[30] See note 3.
[31] *CWL3*, ch. 8, 270-295.

the future philosopher working within the new method, however, it eventually will be normal to reach a control of meaning in representative examples from up-to-date chemistry. Otherwise, talk of "chemical equations," "atomic weights," "molecules," "compounds" and "biomolecules" will be out-of-date or will lack control of meaning in images, descriptions, names and techniques. Implicitly, therefore, the book also is an invitation to learn something of modern chemistry, self-attentively. Similar remarks apply to physics and all of the developing sciences and technologies. I am, though, referring to future foundational progress in the community. In the meantime, the present book invites interest in, and beginnings in, the new method, through discussion, reflection and preliminary self-attention in a range of contexts outlined below.

I.4   Chapters of the Book

For the convenience of readers making use of online search engines, each chapter begins with a detailed abstract. Here, I provide a broad outline of the book chapters and Epilogue.

The *first chapter*, "Space and Time," is "a bridge"[32] into the book. The chapter draws attention to the concreteness and historicity of the Space and Time problem. Lonergan's two main theorems on, respectively, the abstract intelligibility and the concrete intelligibility of Space and Time are discussed in Chapter 5 of the present book. Chapter 1 keeps a focus on introductory problems and ends with a pointer to the fact that part of what is needed are additional elementary examples.

The *second chapter*, "A Foundations Lift from the Adult *Columbidae*," enters avian science.[33] The chapter makes some progress toward intimating the possibility of a verifiable explanatory heuristics of the *adult* pigeon. Keeping discussion to the *adult* pigeon makes it possible to temporarily defer more challenging questions about *organic development*.

---

[32] *CWL3*, 163.

[33] See, for example: Lewis Stevens, *Avian Biochemistry and Molecular Biology*. Cambridge University Press, Cambridge, 1996.

A heuristics reached in Chapter 2 leads to fundamental questions about living organisms, reflections about which follow in later chapters.

The *third chapter*, "Growing to Flight Mastery: The Whole Storeyed Story," looks to the problem of *organic development*. Special attention is given to avian development, one of the most studied in 20th century embryology and developmental biology. A main objective is to do for avian development what Chapter 2 did for the adult bird. That is, Chapter 3 makes some progress toward intimating the possibility of explanatory heuristics for avian development; and then of organic development more generally.

The *fourth chapter*, "Biological Entities," includes a question posed by, among others, Erwin Schrödinger: *What is Life?*[34] In his effort to account for what has sometimes been called a reverse entropy in organisms, Schrödinger initially suggested that the numerical quantity *negative entropy* might be helpful. Soon after, he abandoned that approach.[35] Later, in 1953, entropy with a negative sign was introduced into information theory, and was named *negentropy*.[36] The fourth chapter of the present book does not enter into information theory. Instead, preliminary reflections suggest that, once suitably transposed, Schrödinger's more basic suggestions about entropy and negentropy refer to real aspects of both organic and non-organic entities. The *fourth chapter* also makes progress toward a heuristics of living and non-living entities, *autonomic forms* and *synnomic forms*, respectively.

---

[34] Erwin Schrödinger, *What is Life?: With Mind and Matter and Autobiographical Sketches*, Cambridge University Press, Cambridge, 2003. (First published 1944, *What is life? The Physical Aspect of the Living Cell*. Based on lectures delivered under the auspices of the Dublin Institute for Advanced Studies at Trinity College, Dublin, in February 1943.)

[35] "The remarks on negative entropy have met with doubt and opposition from physicist colleagues. ..... 'entropy taken with a negative sign,' which by the way is not my invention. It happens to be precisely the thing on which Boltzmann's original argument turned. But F. Simon has very pertinently pointed out to me that my simple thermodynamical considerations cannot account for our having to feed on matter 'in the extremely well ordered state of more or less complicated organic compounds' rather than on charcoal or diamond pulp. He is right" (Erwin Schrödinger, "Note to Chapter 6," in *What is Life?*).

[36] Leon Brillouin, (1953) "Negentropy Principle of Information," *Journal of Applied Physics*, vol. 24 (9) (1953): 1152-1163; and Léon Brillouin, *La science et la théorie de l'information*. Paris: Masson, 1959.

The *fifth chapter*, "The Concrete Intelligibility of Space and Time," lifts previous chapters into a larger context. Chapter 5 includes questions about the universe and what Lonergan called *emergent probability*. The chapter ends by raising the problem of implementation. This also serves as segue to the Epilogue.

The *Epilogue*, "What is Science?", picks up on the problem of implementation raised at the end of Chapter 5. The discussion draws attention to a later (1965) discovery of Lonergan, the discovery of an omnidisciplinary methodology (pre-) emergent in history that he called *functional specialization*.[37] Probabilities of emergence and survival of generalized empirical method will increase within the supporting context of an emerging "specialized auxiliary,"[38] that is, beginnings toward functional specialization. And, mature empirical method will be functional. Of course, the Epilogue can only lightly touch on these large and complex issues. However, references to the literature are given. Also, it was important to draw attention to the fact that implementation of the solution to the problem of cultural progress and decline[39] will be intrinsic to adequate empirical method in the sciences.

---

[37] See Epilogue.

[38] *CWL3*, 747.

[39] See, for example, *CWL3*, "General Bias," sec. 7.8, 250-269, and discussion there of cycles of decline in culture.

Chapter 1

Space and Time

Abstract: The first five sections of this chapter draw attention to scientific and philosophic searchings about space, time, extensions and durations. Section 6 looks to Lonergan's solution, a compactly expressed heuristics. By tracking the discussion in *Insight*,[1] it becomes evident that the solution presented in Chapter 5 of *Insight* emerges from an extraordinary control of meaning attained by Lonergan. That chapter of *Insight* includes a precisely expressed two-part heuristics. In one part, Lonergan speaks of the "abstract intelligibility of Space and Time" evidenced in geometric field theories.[2] In a second part, he speaks of the "concrete intelligibility of Space and Time,"[3] "an intelligibility grasped in the totality of concrete extensions and durations"[4]: For the genius Lonergan, "the answer is easily reached"[5] and he named it "emergent probability."[6] As will emerge gradually in this book, adequate heuristics of both the abstract and the concrete intelligibility of Space and Time at the level of the times[7] will be something for the future academy. Modest beginnings toward that goal are to be made with elementary examples. Section 7 is a series of exercises toward describing one's own (re-)discovery of Galileo's law of falling

---

[1] Bernard Lonergan, *Insight, Collected Works of Bernard Lonergan*, vol. 3. Toronto: University of Toronto Press, 1992. Hereafter referred to as *CWL3*, or *Insight*.

[2] Lonergan, *CWL3*, 172-174,

[3] Lonergan, *CWL3*, 194-195.

[4] Lonergan, *CWL3*, 195.

[5] Lonergan, *CWL3*, 195.

[6] Lonergan, *CWL3*, 195. See Section 9.

[7] As Lonergan quoted from Ortega y Gasset, in the original Preface to *Insight*: 'one has to strive to mount to the level of one's times.' See, Frederick E. Crowe, *Lonergan* (St. John's Abbey, Collegeville, MN: The Liturgical Press, 1992), note 1, 58.

bodies. There are various reasons for discussing the Galileo breakthrough, four of which are: 1. the example is part of the tradition of physics and philosophy of space and time; 2. it will be accessible to a wide audience; 3. it provides an elementary exercise for beginnings in the new method; and 4. through self-attention, one can witness the death of the mistaken conflict between so-called primary and secondary qualities. Section 8 is for beginnings in statistical method. A contemporary heuristics of science will need to have a control of meaning in empirical probability, as well as a heuristics of the normative complementarity of classical and statistical methods. Section 9 provides glimpsings of the journey ahead, some details of which will be explored further in the rest of the book.

1.1 Preamble

What are space and time? Space can be described as the volume in which we move about, and the volumes in which spatial and temporal things exist. In a similar way, time can be said to be the duration of our living and moving about, and the duration of things existing in space. As history shows, however, asking about space and time soon leads to difficult questions in science, philosophy, metaphysics and cosmology.

In fact, there is a long history of thought about space and time, and the cosmos.[8] Writings on these questions go back millennia and (as is well known) in more recent times increasingly have relied on mathematics and physics. In physics, advances made by Einstein in Special and then General Relativity prompted fundamental shifts in heuristics of space and time.[9] In gravitation theory, one usually considers large-scale cosmological distance and time scales.[10] In 20th and 21st century particle

---

[8] See, for example, John North, *Cosmos, An Illustrated History of Cosmology and Astronomy*. Chicago: University of Chicago Press, 2008.

[9] However, there is ongoing debate about the significance of Einstein's work. Chapter 5 of *Insight, CWL3*, provides leads.

[10] This is a large and active area of ongoing research. A classic is the text by Hawking and Ellis: S. W. Hawking and G. F. R. Ellis, *The Large-Scale Structure of Space and Time* (Cambridge Monographs in Mathematical Physics). Cambridge: Cambridge University Press, 1973.

physics, distance and time scales are very much smaller.[11] In the 20th century, the physics community gradually worked out what is now known as the Standard Model.[12] The Standard Model is a quantum gauge field theory, and is still surviving as a best-to-date understanding of complex mesh-works of space-time trajectories that, at CERN, for example, are digitally generated screen-images.[13]

In addition to experimental and mathematical physics, there are also philosophic traditions that appeal to contemporary physics.

> Why is there a 'philosophy of space and time?' It seems obvious that any serious study of the nature of space and time, and our knowledge of them, must raise questions of metaphysics and epistemology. It also seems obvious that we should expect to gain some insight into those questions from physics, which does take the structure of space and time, both on small and on cosmic scales, as an essential part of its domain.[14]

There are, though, other kinds of philosophic work about space and time. For example, there is Kant's famous 18th century effort to

comprehend Euclid and Newton within a theory of the synthetic *a priori*.[15]

Whatever one might think about Kant's methods, his ideas about space and time did not survive the advances of either 19th century geometry or 20th century physics. By the end of the 19th century it was known that Euclidean geometry was but one instance among infinite families of mathematical geometries[16]; and in physics, Kant's conjectures about

[11] Mark Thomson, *Modern Particle Physics* (Cambridge: Cambridge University Press, 2013), sec. 2.1, "Units in Particle Physics," 30-32.

[12] Lochlainn O'Raifeartaigh, *The Dawning of Gauge Theory*. Princeton: Princeton University Press, 1997.

[13] In one of the tracking devices used at CERN, computer programs convert aggregates of electrical signals into trajectory images on computer screens. For more detailed descriptions, see "How a detector works," CERN, http://home.web.cern.ch/about/how-detector-works.

[14] Robert DiSalle, *Understanding Space-Time, The Philosophical Development of Physics from Newton to Einstein* (Cambridge, Cambridge University Press, 2006), 1.

[15] DiSalle, x.

[16] Jeremy Gray, *Ideas of space: Euclidean, non-Euclidean, and relativistic*, 2nd ed. Oxford: Clarendon Press, 1989. Martin Jay Greenberg, *Euclidean and non-Euclidean geometries:*

'absolute space' (based partly on his descriptions of water in a spinning bucket) did not contribute to developments in Electrodynamics, Special Relativity and General Relativity, let alone experimental and theoretical results in 20th and 21st century quantum field theories. Kant's work is just one example of what in more recent times increasingly is considered to be an antiquated style of philosophy of space and time. For it is now generally accepted that (a) old fashioned metaphysical discourse about space and time no longer is adequate and (b) that metaphysical arguments are

accountable to the physics of space and time.[17]

However, even in contemporary philosophy of physics, debates continue, partly because there is not yet any consensus on how, precisely, results need to be accountable to physics.[18] The literature on space and time continues to expand with books and articles that appeal to physics not only in diverse ways, but regularly in philosophically incompatible ways.[19]

This is not, however, a merely philosophic problem. And, detailed descriptions by DiSalle[20] tell something of an opposite influence, wherein philosophies of space-time have been affecting developments in experimental and theoretical physics. See DiSalle's book for various examples in the period from Newton to Einstein. For an example from

---

Development and history. New Third edition. York, NY: W. H. Freeman and Co, 1993; B. A. Rosenfeld, *A history of non-Euclidean geometry: evolution of the concept of a geometrical space.* New York: Springer, 1988.

[17] DiSalle, 7.

[18] A masterful discussion of the problem is in, Bernard Lonergan, "A Note on Geometrical Possibility," ch. 6 in *Collection*, vol. 4 of *Collected Works of Bernard Lonergan* (Toronto: University of Toronto Press, 1988/93), 92-107. Eventually, Lonergan's article will need to be read within an increasingly refined generalized empirical method, with data obtained in up-to-date geometric field theories.

[19] For samplings of views, see: (1) Robert Batterman, ed., *The Oxford Handbook of the Philosophy of Physics.* Oxford: Oxford University Press, 2013; (2) Dean Rickles, ed., *The Ashgate Companion to Contemporary Philosophy of Physics.* Burlington, VT: Ashgate Publishing Company, 2008; (3) Jeremy Butterfield and John Earman, eds., *Handbook of the Philosophy of Science, Philosophy of Physics* (Amsterdam: North-Holland (Elsevier), 2007); and (4) John Earman and John D. Norton, *The Cosmos of Science, Essays of Exploration.* Pittsburgh: Pittsburgh University Press, 1997.

[20] DiSalle, 2006.

more recent times, we can look to string theory. String theory (or, in more up-to-date terms, *superstring theory*) is an infinite family of different geometries, heuristically defined. As is well known, efforts within string theory heuristics have led to notable mathematical developments in differential geometry. But, which of the infinite number of possible string theoretic structures might be relevant to physics?[21] Or, if we want a negative cosmological constant, the upper bound on the number of possible theories is of the order 10^{500}.

As large as that number is, though, it is not the number in itself that has been giving some physicists pause. There are fundamental difficulties with string theory heuristics.[22] These difficulties include that, so far: (a) Only relatively few mathematical string theory geometries have been identified; and, in those cases, the mathematical structure is only partially and indirectly known through approximation methods; (b) What predictions have been possible are inconsistent with known experimental results; and (c) Theoretical energy levels that would (in principle) be needed to explore string theory unifications of quantum field theories with gravitational field equations (to potentially produce a *Theory of Everything*), are many orders of magnitude beyond presently known upper limits.

Despite these problems, however, string theory has come to dominate research agendas in contemporary departments of theoretical physics, as well as institutes and funding agencies supporting theoretical physics. Why is this happening? Progress toward answering that question would need to include up-to-date historical analyses. For present purposes (that is, to illustrate an explicit influence of philosophy on physics), it is enough to observe that some of the leading proponents of string theory defend the

---

[21] "If we want to get a negative or zero cosmological constant, there are an infinite number of distinct theories. If we want the theory to give a positive value for the cosmological constant, so as to agree with observation, there are a finite number; at present there is evidence for 10^{500} or so such theories (Lee Smolin, *The Trouble with Physics, The Rise of String Theory, the Fall of a Science, and What Comes Next* (Boston: Houghton Mifflin Co., 2006), 158). See also, Peter Woit, *Not Even Wrong: The Failure of String Theory and the Search for Unity in Physical Law* (New York: Basic Books, 2006), 242. For details on the emergence of string theory, see "A Brief History of String Theory," Part II, chs. 7 -12 in, Lee Smolin, *The Trouble with Physics,* 101-199.

[22] For more details, see references in note 21.

string theory heuristics by explicitly appealing to philosophical views whereby mathematical elegance of string theories is taken to imply physical reality; and that different vacuum states, therefore, represent different universes within a multiverse.[23] For string theorists,

> (t)he long-held hopes for a unique theory have receded, and many (string theorists) now believe that string theory should be understood as a vast landscape of possible theories, each of which governs a different region of a multiple universe.[24]

While string theory has come to dominate theoretical physics, there have been (and are) leading physicists and research groups who in various ways have been doubtful of the string theory agenda. These include Richard Feynman (1918-1988),[25] Lee Smolin,[26] Roger Penrose,[27] Sheldon

---

[23] See, for example, Leonard Susskind, *The Cosmic Landscape: String Theory and the Illusion of Intelligent Design.* New York: Little, Brown and Co., 2005: "It's one thing to argue that theory gives rise to many possibilities for the Laws of Physics, but it's quite another to say that nature actually takes advantage of all the possibilities. Which of the many possible environments materialized as real worlds? ... Discovering that String Theory has 10^{500} solutions explains nothing about our world unless we also understanding how the corresponding environments came into being. ... Is there a natural mechanism that would have populated a megaverse with all possible environments, turning them from mathematical possibilities to physical realities? This is what an increasing number of theoretical physicists believe – myself included. I call this view the *populated Landscape*" (Susskind, 293-4; ff). See also, Lee Smolin, *The Trouble with Physics*, Part II.

[24] Smolin, 149.

[25] "So the fact that it might disagree with experiment is very tenuous, it doesn't produce anything; it has to be excused most of the time. It doesn't look right" (Richard Feynman, secondary source, P.C.S. Davies and Julian Brown, eds., *Superstrings: A Theory of Everything* (Cambridge: Cambridge University Press, 1988), 195).

[26] Lee Smolin, *The Trouble with Physics.*

[27] Penrose, "The physical status of string theory," sec. 31.18 in, Roger Penrose, *The Road to Reality, A Complete Guide to the Laws of the Universe* (New York: Alfred A. Knopf, 2004), 926-929.

Glashow,[28] Steven Weinberg,[29] and many others. And, whatever string theorists may suggest about imagined multiple universes, there are international teams of front-line experimental physicists who diligently struggle in this universe to find potentially significant data in particle accelerators and astronomical observations.

In *The Road to Reality*,[30] Penrose comments on the debate:

> mathematical coherence and elegance, in the mathematics of a physical theory, despite their undoubted value, are clearly far from sufficient.[31]

He also notes that

> there is (as yet) no observational reason to believe that string theory (in particular) *is* physics.[32]

What, though, might "observational reason" mean, or "mathematical coherence and elegance" and "physical theory"? Earlier in his book, he speaks of "three profound mysteries in the connections between" what he calls the "Platonic mathematical world," the "physical (world)" and the "mental (world)."[33] In the concluding section of his book, Penrose outlines further aspects of his searchings, "Deep questions answered, deeper

---

[28] "The string theorists have a theory that appears to be consistent and is very beautiful, very complex, and I don't understand it. It gives a quantum theory of gravity that appears to be consistent but doesn't make any other predictions. That is to say, there ain't no experiment that could be done nor is there any observation that could be made that would say, 'You guys are wrong.' The theory is safe, permanently safe. I ask you, is that a theory of physics or a philosophy?" (Sheldon Glashow, *Viewpoints on String Theory*, Nova Science Programming On Air and Online, http://www.pbs.org/wgbh/nova/elegant/view-glashow.html).

[29] "Though mathematics is used in the formulation of physical theories and in working out their consequences, science is not a branch of mathematics, and scientific theories cannot be deduced by purely mathematical reasoning" (Steven Weinberg, *To Explain the World, The Discovery of Modern Science*. New York: Harper, 2015).

[30] Roger Penrose, *The Road to Reality*.

[31] Roger Penrose, *The Road to Reality*, 1015.

[32] Penrose, 1017. Italics in source text.

[33] Penrose, "Three worlds and three deep mysteries," sec. 1.4, 17-24.

questions posed."[34] Recalling examples of 20th century progress in science and technology, Penrose writes that

> (m)any of these developments depend directly upon physics in one form or another. Moreover, the basic rules of chemistry, as understood today, are also fundamentally physical ones (in principle if not in practice) – mainly coming from the rules of quantum mechanics. Biology is a good deal further from being reducible to physical laws, but we have no reason to believe (consciousness apart[35]) that biological behaviour is not, at root, purely dependent on physical actions that we now basically understand. Accordingly, biology seems to be ultimately controlled by mathematics. ... (A) plant's growth is ultimately controlled by the same physical forces that govern individual particles of which it is composed.[36]

Penrose then expands the context, writing of physical forces in the Earth's magma, volcanic eruptions and effects of such on cultures and communities.[37]

Penrose is one of many contemporary scholars who have been working on basic questions about space, time, cosmology, emergence, the sciences, knowledge, growth and development. In addition to the references given in note 19 (above), one also may look to the essays in, *Science and Ultimate Reality, Quantum Theory, Cosmology and Complexity*.[38] Topics there include: quantum reality, information theory, cosmology, biological life and emergence, consciousness, mind and autonomous agents. The essays there point to real complexities of the universe. Our spatial and temporal universe is all of physics; chemistry; biology; vast aggregates of entities, and ongoing emergence of entities. Whatever *emergence* will mean, it includes also emergence of human organisms who think and write about emergence, science and reality. However, much as in the philosophy of science literature generally, the collection of articles in *Science and Ultimate Reality* does not reveal

---

[34] Penrose, sec. 34.10, 1043-1045.

[35] Roger Penrose, *Shadows of the Mind: A Search for the Missing Science of Consciousness*. Oxford: Oxford University Press, 1994.

[36] Penrose, *The Road to Reality*, 1043.

[37] Penrose, *The Road to Reality*, 1044.

[38] John D. Barrow, Paul C. W. Davies and Charles L. Harper Jr, eds., *Science and Ultimate Reality: Quantum Theory, Cosmology, and Complexity*. Cambridge: Cambridge University Press, 2004.

emergence toward a common view or shared heuristics. Instead, many viewpoints openly are at odds with each other and regularly depend on mutually incompatible philosophic arguments.

This section began with a question: 'What are space and time?' One purpose of this preamble has been to help bring out that even elementary features of the problem ultimately are magnificently challenging. Historical developments reveal that questions about space and time eventually will need to be handled within some combination of all of the sciences and philosophy, within an up-to-date heuristics of an emergent cosmos. The context is large and evidently calls for progress in empirical methods. The next section in this chapter is to help somewhat in preparing the ground for that possibility.

1.2 Preliminary Adjusting

In the Preface, I mentioned that the intended audience of this book includes readers already familiar with Lonergan's work, and especially those who have made some progress toward reaching some version of a *basic position*.[39] But, even if the reader is not familiar with Lonergan's description of *basic position*, every scholar has some orientation about what is real and what is not, what is possible and what is not, about what would be progress, and what would not be progress.

For those not in science and philosophy, this is usually not a problem. But, determining what is real and what is possible are both intrinsic to professional inquiry and not easy to do. The contemporary scientific effort includes results about elementary entities that are represented in diagrams, but that cannot be seen. It also includes inquiry about human consciousness, consciousness that, whatever else, partly is biochemical and partly psychological. Biochemistry and human psychology are known to be integral to the possibility of human intellectual activity. The contemporary philosophical effort includes inquiry into all of this, including inquiry into biochemical-psychological-intellectual activity that is scientific thinking about non-visible elementary physical entities. Adding to the challenge, there is the possibility of adverting to one's own

---

[39] *CWL3*, 413. See also Section 2, below.

operations, and that way reaching preliminary descriptions of our dynamics of knowing and doing.[40] Or, one may at least observe that understanding is not reached without some kind of sense experience; that understanding is reached by a focusing of attention on patterned sounds, sights, touch (e.g. Braille), written words or symbols, diagrams; and that all of these experiences are within the fabric of one's own biological consciousness.[41]

Whether imagining electrons or quarks, reading linguistic or scientific symbolisms, or inspecting patterned trajectories from a digital screen in a laboratory, all of these experiences have something in common: they all can be described as being partly spatial and temporal. Now, it is not at all new to observe that space and time are fundamental to human experience and knowledge. How, though, are we to make progress toward a heuristics at the level of the times?

> Not only are these notions (of space and time) puzzling and so interesting, but they throw considerable light on the precise nature of abstraction. They provide a concrete and familiar context for … analyses of empirical science, and they form a

---

[40] The possibility is pointed to in the diagram in *CWL3*, 299; and *CWL3*, sec. 18.2.2 – 18.2.5, 632-639. See also, Bernard Lonergan, *Phenomenology and Logic, The Boston Lectures on Mathematical Logic and Existentialism*, vol. 18 in, *Collected Works of Bernard Lonergan*, ed. by Philip J. McShane (Toronto: University of Toronto Press, 2001), "Appendix A, Two Diagrams," 319-323.

[41] Aristotle, Aquinas and Lonergan made similar observations. In Aristotle's *De Anima*, III, 7: "The faculty of thinking then thinks the forms in the images" (Aristotle, *On the Soul* (*De Anima*), tr. by J. A. Smith, Classics in the History of Psychology, http://classics.mit.edu/Aristotle/soul.3.iii.html, in The Internet Classics Archive, http://classics.mit.edu/index.html.). Centuries later, St. Thomas Aquinas, a student of Aristotle's works, observed that: "it is impossible for our intellect to understand anything actually, except by turning to the phantasms.... anyone can experience this of himself, that when he tries to understand something, he forms certain phantasms to serve him by way of examples, in which as it were he examines what he is desirous of understanding" (*Summa Theologiae* I, tr. by Fathers of the English Dominican Province (Benziger Bros. edition, 1947) q. 84, a. 7. Note, too, that the quotation from Aristotle is on the title page of *Insight*. See also, *CWL3*, sec. 1.2.3, "The Image," 33: "(T)he image is necessary for the insight." See also, "Insight into phantasm," sec. 1.4 in Bernard Lonergan, *Verbum: Word and Idea in Aquinas*, vol. 4 in *Collected Works of Bernard Lonergan* (Toronto: University of Toronto Press, 1997 (first published: Notre Dame, IN: University of Notre Dame Press, 1967)), 38-45.

natural bridge over which we may advance from our examination of science to an examination of common sense.[42]

It would be more correct to say that there is the *potential* for such notions to throw light on the nature of abstraction, and empirical science, for those who have been reading *Insight* (or a comparable series of exercises) the way Lonergan asked at the end of Chapter 1 of *Insight*. Toward that possibility, there is the present task of adjusting our present position. I note, here, that a dictionary meaning of 'adjust' includes "to put (e.g., a musical instrument) in good working order, to bring into a proper state or position." How, then, can we adjust our own position so that we become better self-tuned to the problem?

I invite attention to two words that appear in Lonergan's discussion of Space and Time in Chapter 5 of *Insight*, namely, *extension* and *duration*. While the Space-Time problem is perhaps not generally familiar in all quarters of the academic community,[43] the words *extension* and *duration* refer to familiar experiences. And, an effort toward identifying those familiar experiences can help us improve our positioning.

1.3 "All Positions Invite Development"[44]

In Chapter 14 of *Insight* we are invited to what Lonergan calls a *basic position*:

> It will be a basic position (1) if the real is the concrete universe of being and not a subdivision of the 'already out there now'; (2) if the subject becomes known when it affirms itself intelligently and reasonably and so is not known yet in any prior 'existential' state; and (3) if objectivity is conceived as a consequence of intelligent inquiry and critical reflection, and not as a property of vital anticipation, extroversion, and satisfaction.[45]

For those readers not already familiar with the expression 'already out there now,' see the Index of the book *Insight* for numerous references to

---

[42] *CWL3*, 163.
[43] *CWL3*, 163, first paragraph.
[44] *CWL3*, 413.
[45] *CWL3*, 413.

discussion regarding its application and meaning.[46] In fact, Lonergan discusses the expression term by term:

> 'Already' refers to the orientation and dynamic anticipation of biological consciousness; such consciousness does not create but finds its environment; it finds it as already constituted, already offering opportunities, already issuing challenges. 'Out' refers to the extroversion of a consciousness that is aware, not of its own ground, but of objects distinct from itself. 'There' and 'now' indicate the spatial and temporal determinations of extroverted consciousness. 'Real,' finally, is a subdivision within the field of the 'already out there now': part of that is mere appearance; but part is real; and its reality consists in its relevance to biological success or failure, pleasure or pain.[47]

After reading as far as Chapter 14 in *Insight*, the described basic position, as well as the description of 'already out there now real,' can seem fairly reasonable. For some, a preliminary basic position may well already be attained. But, let us push the issue somewhat further, or rather make the effort to push ourselves somewhat further, by asking a perhaps surprising question: Within a basic position, what is it to see color? Within the heuristics of a basic position, we can look forward to an understanding of colors through advances in physics, as well as advances in biophysics and the human faculty of sight. It is well known in contemporary neuroscience that the human organism's sensitivity to light-spectra partly is through photo-receptor cells and cellular pathways which penetrate deep into the brain. But, the sense organ for sight, the brain, and its various tissues and cell growth are spatial and temporal. And, indeed, so are the countless photos, images, and biochemical and biophysical symbolisms one attends to in order to understand the biophysics and neuroscience of human photosensitivity. Within a basic position, then, even if one's

---

[46] See, *CWL3*, Index, under 'real.'

[47] *CWL3*, 276-277. See, also, Bernard Lonergan, *Method in Theology* (London: Darton, Longman & Todd, 1973), 263. Another description is found in Lonergan's later book on method, in a discussion about the world of immediacy: "(T)he object of the world of immediacy ... is already, out, there, now, real. It is *already*: for it is the object of extraverted consciousness. It is *there*: as sense organs, so too sensed objects are spatial. It is *now*: for the time of sensing runs along with the time of what is sensed. It is *real*: for it is bound up with one's living and acting and so must be just as real as they are" (Bernard Lonergan, *Method in Theology*, 263).

inquiry is about color, working toward a heuristics of color one is called to work out heuristics of a range of related experiences, including extensions and durations.

As one may soon find, inquiry about extension and duration raises even more questions. We each have a body and in one way or another we are aware of an extensiveness of our body - our limbs, our hands, our feet. Think of where you are while you are reading this chapter. There is the space of a room, a garden perhaps, or a library, or a seat in a place away from the busyness of a town. In some unfortunate cases, a person may be confined to a bed, or a room. But even then, one is confined to the space of the bed or the room, and even that confinement also implies what would otherwise be possible. For, normally, we move about our home, our town, and we can even imagine the volume or space surrounding our planet. Thanks to photographs and imaging technology from modern telescopes and space probes like the NASA *Kepler*,[48] we also can imagine volumes of space containing distant planets, stars, exotic galaxies and much more besides. The volumes of space that we see and imagine around us seem to go on without any apparent limits. As soon as we imagine some kind of box or volume that might contain one or some of these many objects, we find that we can immediately imagine a larger box. But what, then, could be meant by the first part of the definition of 'a basic position,' that "the real is not a subdivision of the 'already out there now'"?

For, whatever color any of these objects might have, do not our bodies and all of these objects all have lengths, depths, widths, and occupy volumes of space? Is there not, perhaps, some sense in which such objects are, in fact, 'out there' – and for galaxies, '*way* out there'? Such then is the puzzle here: How can we reconcile experience of extension and duration with the assertion that

the real is ... not a subdivision of the 'already out there now'[49]?

These questions invite us to further development.

---

[48]National Aeronautics and Space Administration (NASA), http://kepler.nasa.gov/. Thanks to the recent NASA New Horizons Mission, there are now high resolution photos of the surface of Pluto. https://www.nasa.gov/mission_pages/newhorizons/main/index. html.
[49] See note 45.

1.4 "The Enigma of the Body"[50]

Part of the challenge of the new more balanced empirical method is to attend to all data. Reflections of Merleau-Ponty helpfully deepen the puzzle by inviting attention to data found within the intimacy of one's own experience. His paper "Eye and Mind"[51] begins with a mood-setting quotation:

> What I am trying to convey to you is more mysterious: it is entwined in the very roots of being, in the impalpable source of sensations.[52]

A few pages into his article he writes:

> The enigma derives from the fact that my body simultaneously sees and is seen. ... Visible and mobile, my body is a thing among things; it is one of them. It is caught in the world, and its cohesion is that of a thing. But because it moves itself and sees, it holds things in a circle around itself. ... These reversals, ..., are different ways of saying that vision is caught or comes to be in things ... the visible undertakes to see, becomes visible to itself, ..., like the original solution still present within crystal, the undividedness of the sensing and the sensed.[53]

We may look at a tree, its colors and shadows, yes, and we may touch textured lengths of trunk and branch, and see limbs reaching into a spaciousness of foliage – all of this, and we say, "a tree is there." There is

---

[50] Merleau-Ponty, Maurice, "Eye and Mind," sec. II, Original: *L'Œil et l'esprit* (Paris: Gallimard, 1961), par. 10. A translated version: Tr. by Carleton Dallery, in *The Primacy of Perception*, ed. by James Edie (Evanston: Northwestern University Press, 1964), 159-190. Revised translation by Michael Smith in *The Merleau-Ponty Aesthetics Reader* (1993), 121-149. For "Eye and Mind," see also, http://www.biolinguagem.com/biolinguagem_antropologia/merleauponty_1964_eyeandmind.pdf.

[51] Merleau-Ponty, "Eye and Mind."

[52] A quotation from Paul Cézanne, recorded by: J. Gasquet, *Cézanne* (La Versanne, France: Encre Marine, 2002. Originally published in 1921), 3. Gasquet was an early 20th century commentator on the work of Cézanne.

[53] *The Merleau-Ponty Aesthetics Reader*, 125.

a locality, from which height, width, and depth are abstracted, a voluminosity we express in a word when we say that a thing is there.[54]

Do these reflections not add to the existential puzzle? Merleau-Ponty certainly seems to be getting at something, perhaps something that we too can find in our experience. Are we not aware, at times, of objects at hand (or at foot), that have length, depth, volume, that occupy space, that in some basic way would seem in fact to be 'already out there now'? Does there not seem to be some kind of bodily knowledge of "voluminous" or extended things "there"? In some version of, or approximation to 'a basic position,'[55] how might we account for bodily experiences named extension and duration?

1.5 Primary Qualities and Secondary Qualities[56]

Another approach to questions about experience and objectivity is found in a philosophic tradition that includes Descartes (1596–1650), Galileo (1564–1642), Hobbes (1588–1679), Boyle (1627–1691), Locke (1632–1704), Berkeley (1685–1753), Kant (1724–1804) — and others up to the present day. This tradition also draws attention to experience, but in a different way with different results.

An early contribution to this tradition was Galileo's 16th century discovery of the Law of Falling Bodies. Galileo reached an understanding of free-fall in terms of a quadratic equation by which one correlates measured distances and measured times.[57] But, whether expressed with imagined similar triangles, mathematical symbols or Italian words,[58] the

---

[54] Merleau-Ponty, *Eye and Mind*, 13.

[55] *CWL3*, 413.

[56] References in *CWL3*: 107-109, 123, 153, (177-178), 277, 319, 363, 438 (-440). Additional references added, but not found in the Index of *CWL3*, are indicated by parentheses.

[57] Galileo Galilei, *Dialogues Concerning Two New Sciences* [1638], trans. Henry Crew and Alfonso De Salvio (New York: McGraw-Hill Book Company, 1963), "Naturally Accelerated Motion." Online, see, for example, Galileo Galilei, *Dialogues Concerning Two New Sciences* [1638], Online Library of Liberty, http://oll.libertyfund.org/.

[58] Galileo Galilei, *Dialogues Concerning Two New Sciences* [1638]. Galileo wrote his book in Italian.

correlation he discovered evidently is nothing like what one sees when observing metal spheres rolling down inclined planes or cannon balls in flight. Yet, by making use of apparatus available at the time, Galileo's equations seemed to be verified.[59] But, what then, of colours, taste, smell and sound? How might all of these various differences be accounted for? Galileo advocated two allegedly complementary notions[60] that subsequently were taken up by the philosophic tradition mentioned above:

> Whereas *primary qualities*—such as figure, quantity, and motion—are genuine properties of things and are knowable by mathematics, *secondary qualities*—such as colour, odour, taste, and sound—exist only in human consciousness and are not part of the objects to which they are normally attributed.[61]

---

[59] See notes 77, 78 and 79; and 57. Galileo Galilei, "Naturally Accelerated Motion," discussion following Corollary I of Theorem II, Proposition II.

[60] "Now I say that whenever I conceive any material or corporeal substance, I immediately feel the need to think of it as bounded, and as having this or that shape; as being large or small in relation to other things, and in some specific place at any given time; as being in motion or at rest; as touching or not touching some other body; and as being one in number, or few, or many. From these conditions I cannot separate such a substance by any stretch of my imagination. But that it must be white or red, bitter or sweet, noisy or silent, and of sweet or foul odour, my mind does not feel compelled to bring in as necessary accompaniments. Without the senses as our guides, reason or imagination unaided would probably never arrive at qualities like these. Hence I think that tastes, odors, colors, and so on are no more than mere names so far as the object in which we place them is concerned, and that they reside only in the consciousness. Hence, if the living creature were removed, all these qualities would be wiped away and annihilated. But since we have imposed upon them special names, distinct from those of the other and real qualities mentioned previously, we wish to believe that they really exist as actually different from those. ... To excite in us tastes, odors, and sounds I believe that nothing is required in external bodies except shapes, numbers, and slow or rapid movements. I think that if ears, tongues, and noses were removed, shapes and numbers and motions would remain, but not odors or tastes or sounds. The latter I believe are nothing more than names when separated from living beings, just as tickling and titillation are nothing but names in the absence of such things as noses and armpits" (Galileo Galilei, *The Assayer*, 1623, trans. Stillman Drake, in *Discoveries and Opinions of Galileo* (New York: Doubleday, 1957), 274-277).

[61] *Encyclopedia Britannica*, epistemology (philosophy): Epistemology and modern science; http://www.britannica.com/EBchecked/topic/476194/primary-quality.

From this perspective, imaginable lengths, widths, depths, volumes - matter in motion - are real and objective while other secondary qualities are merely apparent. Lonergan suggests that this was a

mistaken twist given to scientific method at the Renaissance.[62]

Whatever Lonergan may have thought, how do primary and secondary qualities fit with *your* present positioning? Contemporary neuroscience is making impressive progress explaining human sight, smell, taste and hearing. Contemporary results partly come from *systems neuroscience* which includes, for example, mathematical geometries of biochemical as well as neural networks, with corresponding statistics and probabilities.[63] What, then, might be the primary and second secondary qualities of human sight, smell, taste and hearing? Which of these contemporary results reveal primary qualities, and which reveal secondary qualities?[64]

Further aspects of the problem come to light when, in a basic position, one asserts that

the real is ... not a subdivision of the 'already out there now.'[65]

What might be the basis, then, for the assertion that dimensions of a mathematical equation represent real and imaginable matter in motion? Within your present position, how does the notion of secondary quality sit with you, the notion that much of sense experience is to be regarded as

---

[62] *CWL3*, 107. See also *Method*, Sec. 10.3, "The Second and Third Stages of Meaning," 93: "In the second stage the world mediated by sense splits into the realm of common sense and the realm of theory."

[63] The literature is vast. See, for example, Daniel H. Geschwind and Genevieve Konopka, "Review Article. Neuroscience in the era of functional genomics and systems biology," *Nature* 461 (October 15, 2009): 908-915; and N. Le Novère, ed., *Computational Systems Neurobiology*. Dordrecth: Springer, 2012. For an example, authors of one article in the N. Le Novère collection write: "(I)f we have any hope of building truly predictive models, the intricate geometries and large length scales of neurons compel us to explicitly account for molecular diffusion and spatial organization" (Sherry-Ann Brown, Raquel M. Holmes and Leslie M. Loew, "Spatial Organization and Diffusion in Neuronal Signaling," ch. 5 in, N. Le Novère, N. ed., *Computational Systems Neurobiology*, 133).

[64] This issue comes up again later, in the fuller context of Chapter 4.

[65] See note 47.

merely apparent? Obviously, there is something amiss here. But, again, how might we account for these differences and begin to resolve these difficulties?

1.6 Lonergan's Solution to the Extension and Duration Problem

Once again, some help can be had by following up on Merleau-Ponty's reflections, where a key part of the exercise is to reflect on one's own experience:

> I must acknowledge that the table before me sustains a singular relation with my eyes and my body: I see it only if it is within their radius of action; ... What is more, my movements and the movements of my eyes make the world vibrate - ... I would express what takes place badly indeed in saying that here a 'subjective component' or a 'corporeal constituent' comes to cover over the things themselves: it is not a matter of another layer or a veil that would have come to pose itself between them and me.[66]

Taking Merleau-Ponty's example, a table, you might reach out and touch the edge of a table at hand (or, even your own other hand, at table). You might feel the warmth of wood, or the cooler feel of some other material. A table may have a color of wood, or be painted a bright red. In touch, or sight, or both simultaneously, we find that edges of a table (or even the fingers and palm of a hand) have what we call length, depth and volume.

Where, though, are these lengths and volumes? Merleau-Ponty's reflections can encourage one to self-attention. Are not all of these experiences, experiences in one's own sense of touch and one's own sense of sight? There can be a seen extension of a table; and there can be a tactile sense of length, depth, volume. Yet, whether experienced separately or seamlessly blended, are not all of these in one's own sense experience? And might, then, extension and duration eventually be grasped as the same kind of experience as color, taste, warmth, sound, scent, or any other

---

[66] Merleau-Ponty, Maurice, *The Visible and the Invisible* (Evanston, IL: Northwestern University Press, 1968), 7.

experience within sensitive consciousness? In that way, the peculiar status of "primary qualities" would die: the death, though, is best witnessed internally by the geometry of space and time.

This brings us to the edge of Lonergan's solution. However, while elegant, the solution he reaches is not easy. In his book *Insight*,[67] Lonergan invites us to climb up through a graduated series of displacements - in experience, understanding, and self-attention. In his solution, colors, extension and duration, or any other sense experience - experienced, remembered or imagined – are all within the sensitive psyche, and all such experience provides data for inquiry (hintings toward the full relevance of the Canon of Complete Explanation[68]). His solution not only depends on a gradually increasing range of contexts, but is open to, and calls for further adjustments and developments from initial displacements. Eventually, all data are subjective, and yet no data are merely subjective. In other words, there is the invitation to make progress within a generalized empirical method.

Relevant here are Merleau-Ponty's subtle descriptions and struggles about a "reversal."[69] Lonergan went on to luminously subsume extension and duration along with color and other sense experience. With a nuanced heuristics of an advanced basic position he was able to speak in precise terms of a

> reversal of roles, in which the sensible container becomes the intellectually contained. ... So it comes about that the extroverted subject visualizing extensions and experiencing duration gives place to the subject orientated to the objective of the unrestricted desire to know and affirming beings differentiated by certain conjugate potencies, forms, and acts grounding certain laws and frequencies.[70]

However, recalling the words of Lonergan's solution is not acquiring the solution for ourselves. Still, Lonergan's achievement does suggest that generalized empirical method can yield important results. And, drawing

---

[67] *CWL3*.

[68] *CWL3*, 107. The canon of complete explanation: "The goal of empirical method ... the complete explanation of all phenomena." See also, "the enveloping world of sense," *CWL3*, 559.

[69] See Note 50.

[70] *CWL3*, ch. 16, 537. See also Chapter 4, below.

attention to Lonergan's solution can help motivate one to follow up in the invitation to work toward an improved method. In the present context, it will help us continue with beginnings if we go to an example that was a source of the puzzle about "primary and secondary qualities." In other words, with beginnings in self-attention, we may enquire into understanding and verification of Galileo's discovery, the Law of Falling Bodies.

As mentioned in the Abstract for this chapter, there are various reasons why pausing to reflect on the Galilean breakthrough in physics will be worthwhile. However, looking to Galileo's result will not be sufficient to handle issues about Space and Time in any up-to-date fashion. There have been more than three centuries of development since Galileo's breakthrough. And, within generalized empirical method, even

> the old philosophic opinion that extension is a real and objective primary quality cannot dispense one from the task of determining empirically the correct geometry of experienced extensions and durations.[71]

Also, there are important recent philosophical reflections that also will need to be subsumed within the new empirical method. For instance, in addition to positive advances already present in philosophy of physics tradition, there is the work of Renaud Barbaras. His book, *The Being of the Phenomenon, Merleau-Ponty's Ontology*, includes a complex weave of results - partly historical, partly dialectical, but mainly (descriptive) interpretation of the life-work of Merleau-Ponty.[72] The expertise in self-

---

[71] *CWL3*, 93.

[72] Renaud Barbaras, *The Being of the Phenomenon, Merleau-Ponty's Ontology*. Trans. Ted Toadvine and Leonard Lawlor, Bloomington and Indianapolis: Indiana University Press, 2004 (orig. French version, Editions Jérôme Millon, 1991). "As for the interpretation of the evolution of Merleau-Ponty's thought as a whole, I believe that I can maintain my developmental hypothesis starting from the problem of ideality and truth, that is, from the necessity of passing from a phenomenology of perception – open to the reproach of being nothing other than a psychology of perception – to a *philosophy* of perception, discovering in perception a mode of being that holds good for every possible being" (Barbaras, xxi). "Our intention is neither to take up the movement of (Merleau-Ponty's) thought from one end to the other in order to grasp *The Invisible and the Invisible* (Maurice Merleau-Ponty, *The Visible and the Invisible, Followed by Working Notes*. Evanston, IL: Northwestern University Press, 1968. Trans. by Alphonso Lingis, ed. Claude Lefort, original French

attention and reflection practiced and advocated by Merleau-Ponty, and then further developed by Barbaras,[73] are not mainstream in contemporary philosophy of physics. But, data to which they both appeal certainly will need to be accounted for within an up-to-date generalized empirical method.

1.7 Revisiting Galileo's Discovery: Farewell to Primary and Secondary Qualities

If a ball is dropped from a height, it moves toward the ground. And, it is easy to see that its speed increases. It is said that there is *acceleration.*[74] But, for the physicist, what are *speed* and *acceleration*? And in what way, precisely, is the speed of a free-falling object accelerated?[75]

Let us recall and re-enact some of Galileo's discoveries, and along the way make a few elementary observations about what we are doing. First,

---

version, *Le Visible et l'invisible*. Paris: Editions Gallimard (1964)) as its end point, nor to evaluate his heritage as developed elsewhere. Our intention is to reconceive Merleau-Ponty's ontology on its own terms – which is, in our eyes, the key to this entire thought as well as to numerous contemporary reflections" (Renaud Barbaras, *The Being of the Phenomenon*, xxxi).

[73] Renaud Barbaras, *Desire and Distance, Introduction to the Phenomenology of Perception*. Translated by Paul B. Milan. Stanford: Stanford University Press, 2006.

[74] 1520's, from Latin *acceleratus*, past participle of *accelerare* "to hasten, to quicken," from *ad-* "to" + *celerare* "hasten," from *celer* "swift."

[75] See the last paragraph of Section 5. It is Galileo's elementary discovery that is being reviewed here. But, for later classical physics, *velocity* and *acceleration* are defined as first and second derivatives. "What does a (classical) physicist mean by velocity? He means ds/dt. What does he mean by an acceleration? He means d^2s/dt^2. If you know what is meant by those symbols from the differential calculus, you know exactly what is meant by acceleration and velocity, and if you do not know what those symbols mean, you do not understand acceleration and velocity" (Bernard Lonergan, *Topics in Education, Collected Works of Bernard Lonergan*, vol. 10, ed. by R. M. Doran and F. E. Crowe (Toronto: University of Toronto Press, 1993) 145). For an introduction to elementary calculus, see Terrance J. Quinn, "The Calculus Campaign," *Journal of Macrodynamic Analysis*, 2 (2002): 8-36. In more contemporary gauge field theories, velocity and acceleration terms still are given by first-order and second-order terms, respectively, but using velocity vectors and a connection operator.

let's begin with Galileo's description of the apparatus he used in his experiments.

> A piece of wooden moulding or scantling, about 12 cubits[76] long [about 7 m], half a cubit [about 30 cm] wide and three finger-breadths [about 5 cm] thick, was taken; on its edge was cut a channel a little more than one finger in breadth; having made this groove very straight, smooth, and polished, and having lined it with parchment, also as smooth and polished as possible, we rolled along it a hard, smooth, and very round bronze ball.[77]

What is *speed*? Galileo considered the ratios of distance traversed to time elapsed. But, how might such a ratio be obtained? It is one thing to measure distance. How might time be measured?

> For the measurement of time, we employed a large vessel of water placed in an elevated position; to the bottom of this vessel was soldered a pipe of small diameter giving a thin jet of water, which we collected in a small glass during the time of each descent... the water thus collected was weighed, after each descent, on a very accurate balance; the difference and ratios of these weights gave us the differences and ratios of the times...[78]

Galileo also described details of his experiment.

> Having placed this board in a sloping position, by lifting one end some one or two cubits above the other, we rolled the ball, as I was just saying, along the channel, noting, in a manner presently to be described, the time required to make the descent. We ... now rolled the ball only one-quarter the length of the channel; and having measured the time of its descent, we found it precisely one-half of the former. Next we tried other distances, comparing the time for the whole length with that for the half, or with that for two-thirds, or three-fourths, or indeed for any fraction; in such experiments, repeated a full hundred times, we always found that the spaces

---

[76] A *cubit*: ancient measurement, elbow to finger tips, ancient unit of measure based on the forearm from elbow to fingertip, usually from 18 to 22 inches, early 14c., from Latin cubitum "the elbow," from PIE keu(b)- "to bend." Such a measure, known by a word meaning "forearm" or the like, was known to many peoples (e.g. Greek pekhys, Hebrew ammah, English ell).

[77] Galileo Galilei, *Dialogues*. See note 57.

[78] Galileo Galilei, *Dialogues*. See note 57.

traversed were to each other as the squares of the times, and this was true for all inclinations of the plane, i.e., of the channel, along which we rolled the ball.[79]

Provided in the table below[80] are some of Galileo's measured times and distances. Galileo measured shorter distances with a unit called *points*, which today is about 29/30 mm. As shown on the chart, in total, the ball rolled about 2104 points, or 2104 x 29/30 mm, or about 2,034 mm (2.34 m), less than 7 feet in eight time intervals.

Table 1.1. One of Galileo's experiments: times, distances and ratios.

Time	Distance (points)	Distance divided by 33
1	33	1
2	130	3.94
3	298	9.03
4	526	15.94
5	824	24.97
6	1,192	36.12
7	1,620	49.09
8	2,104	63.76

In one column we have the measurements for time: 1, 2, 3, ..., 8. How far does the ball roll in the first unit of time? In the right column, this distance is labeled 1. What about the measured distance for the second unit of time? In *points*, the distance is 130. But, compared to the first distance over the first unit of time, it is 3.94 x 33, that is, 3.94 times the first distance. And so on. Now, what is each measured distance compared to

[79] Galileo Galilei, *Dialogues Concerning Two New Sciences* [1638]. See note 57.
[80] Some of Galileo's original data and notebooks may be found translated and discussed in the works of Stillman Drake. A convenient secondary source for this particular table of results is: D.W. MacDougal, *Newton's Gravity: An Introductory Guide to the Mechanics of the Universe, Undergraduate Lecture Notes in Physics* (New York: Springer, 2012), 20. Drake's works include details on various ways that Galileo measured distances and times. See, for example, Stillman Drake, *Galileo at Work: His Scientific Biography*. Chicago: University of Chicago Press, 1981. See also, note 1, p. 18 of D.W. MacDougal, *Newton's Gravity*, for further historical references.

the measured distance traversed in the first unit of time (that is, if all the measured distances are divided by 33)? The resulting ratios 1, 3.94, 9.03, … are (approximately) equal to the squares of the measured times 1, 2, 3, …!

This is just one sample from Galileo's work. He went on to an extensive study of free-fall and projectile motion, and much more. His *opera omnia* includes, among other things, detailed drawings, graphical representations, and complex combinations of triangular and parabolic figures from classical geometry.[81] Here, though, instead of moving into those further results, let's pause and, within a mode of elementary self-attention, take (self-) note of a few features of what we obtain in the results just described.

We have measured distances and measured times. Let us start with distances. How are measured distances obtained? There is no attempt here to invoke a theory of measurement, or to reach general results about measurement. This is a preliminary exercise in self-attention, in describing - or rather self-describing - what is happening when obtaining measurements with an elementary ruler. A ruler may be a length of wood, say, with markings that one can see (or, for the blind, touch). It will help to get an actual ruler. Markings also are numbered: 1, 2, 3, … . How, though, are locations of markings and their numberings determined?

One has some kind of elementary length - a finger's breadth, or perhaps a small length of wood such as what in Galileo's time was called a "point" (or for a larger measuring beam, a *cubit*, a unit length that is the distance from one's elbow to one's finger tips[82]). To make a ruler, one places one end of the unit length at the beginning of the ruler; a mark is made where the unit length ends. Then, continue in this way, until a series of markings is produced. Along the now marked ruler, we have visible lengths that we correlate with markings.

Note, here, that there are no numbers yet. How do we go on to include numbers, such as when one measures, say, "5 points"? We do something more here, yes? Do we not also correlate the number '5' with one of the markings, which, as above, one also correlates with a visible multiple of a

---

[81] See, for example, Galileo Galilei, *Dialogues*, note 57.
[82] See note 76.

seen length along the ruler? In reaching elementary measurements there is, then, a nesting of various insights. In saying "5 points," one correlates '5' with what one already is correlating. And, a table of such measurements - such as the second column in the table obtained by Galileo - represents a series of correlations of correlations. This, of course, is only the most elementary self-description. But, let us go on.

What about time measurements? Again, the aim here is not to enter into discussion about theories of time measurement. The challenge is to describe how one measures time in the law of falling bodies. Keeping to that context, we may recall that Galileo used a water-clock to measure time,[83] weighing volumes of water that flowed from a vessel of water, and that water

collected was weighed, after each descent, on a very accurate balance.[84]

All along there are assumptions about rates of flow, as well as a relationship between volume and weight.[85] The exercise is somewhat easier if, instead of following Galileo, one uses a modern graduated cylinder with markings at regular heights along its edge (corresponding to measured equal increments of volume), or some kind of elementary spring or pendulum. In any case, it is through further exercises in self-attention that one can find that, just as for distance measurements, series of time measurements are obtained through series of correlations of correlations.

What, now, of Galileo's insight pointed to above — that the ratios 1, 3.94, 9.03, ... are (approximately) equal to the squares of the measured times 1, 2, 3, ...? In that insight, do we not, in fact, reach yet a further correlation? In other words, in understanding Galileo's Law of Falling Bodies, one reaches a correlation of correlations of correlations.[86]

---

[83] See note 78.

[84] See note 78.

[85] Galileo was aware of Archimedes' principle of displacement (Laura Fermi and Gilberta Bernardini, *Galileo and the Scientific Revolution* (Mineola NY: Dover Pubs., 200; first published, New York: Basic Books, 1961) 22, 23, 69, 114, 118).

[86] Through such exercises, one can take a few steps toward contemporary contexts: "To employ an explanatory conjugate is to turn attention away from all directly perceptible aspects and direct it to a non-imaginable term that can be reached only through a series of correlations of correlations of correlations" (*CWL3*, 271).

Recall, now, that curious distinction made between allegedly primary and allegedly secondary qualities:

> Whereas *primary qualities*—such as figure, quantity, and motion—are genuine properties of things and are knowable by mathematics, *secondary qualities*—such as colour, odour, taste, and sound—exist only in human consciousness and are not part of the objects to which they are normally attributed.[87]

Where, though, are the figures, quantities, motions, measurements, insights and correlations? Where are the contents leading up to, and including, our grasp of the law of falling bodies?

Even though observations here are only the most elementary beginnings toward a new method,[88] do not our efforts already begin to (self-) reveal something about what is going on, indeed, about our what-ing that is going on? Are not geometric images self-evidently images in one's own imagination? Is not what one sees in one's own sight? Are not understandings reached — for example, correlations, correlations of correlations, and correlations of correlations of correlations — within one's understanding? What basis is there, then, for a special status to be given to "figure, quantity and motion,"[89] when "figure, quantity and motion" all are within one's imagination, one's senses and one's understanding?

As mentioned at the end of Section 6, this does not bring us into contact with up-to-date heuristics of space and time, nor are our results here progress enough to solve fundamental problems in contemporary philosophy of science. But, at least the mistake about primary and secondary qualities can begin to pass away. At the same time, one can obtain a first foothold in the long climb toward an up-to-date basic position.[90]

---

[87] See note 61.

[88] Lonergan, *Third Collection*, 141.

[89] See notes 61 and 87.

[90] *CWL3*, 413. See also note 70.

1.8   June Drop Apples and Empirical Probability

In the first five chapters of *Insight*,[91] Lonergan invites the reader to climb a steepening series of exercises in self-attention. Strategically selected examples are drawn from across more than 21 centuries[92] years of development in mathematics and physics. Even the first example on the first page of the book was, at least for the discerning Lonergan, a "Dramatic Instance."[93] In that first page he invites the reader to also enjoy a similar experience — by re-discovering, and attending to one's re-discovering of the principle of displacement originally discovered by Archimedes. By chapters four and five of the book, however, Lonergan's invitation is given from altitudes of 20th century mathematics and physics, a context that includes statistical mechanics, quantum theory, special relativity and general relativity. In fact, his invitation already goes much further than mathematics and physics, when he speaks of

conditioned schemes of recurrence[94]

and the totality that he called *emergent probability*.[95] However, in our present efforts to make beginnings in a new method, for now, let us look to a single question: What is empirical probability?

The question may seem simple. However, it is not. And, statistical results and problems of interpretation of such occur in complex layerings in contemporary mathematics, physics, biology and philosophy of science. Even if so far mainly only implicit, the need and possibility of control of

---

[91] *CWL3*.

[92] For Lonergan, a convenient source for physics of the time was R. B. Lindsay and H. Margenau, *Foundations of Modern Physics*. New York: Dover Publications, 1957; first published in 1936, first Preface completed in 1935. Archimedes' dates are circa 287–212 B.C.E. Not needing extreme precision here, as an approximation for the number of years since Archimedes' work, 1935 years (A.D.) plus 212 years (B.C.E.) is 2147 years.

[93] *CWL3*, sec. 1.1, 27.

[94] *CWL3*, sec. 4.2.2, 141.

[95] *CWL3*, 144.

meaning within the contemporary context slowly is emerging.[96] The
problem for us is two-fold: one needs to have one's own examples in
understanding probabilities; and there is the (new) task of self-attention.
However, even making beginnings here has its special challenges. One
needs at least some understandings of empirical probabilities. But,
reaching such understandings takes some work. Originally,

> (t)he necessary mathematics all developed from the fundamental principles of
> mathematical probability laid down by Fermat and Pascal in about three months by
> a painstaking application of uncommon sense.[97]

And, both Fermat[98] and Pascal[99] were highly skilled 17th century
mathematicians.

For the present section, we can, instead, move to a more elementary
context. In fact, we can again take help from Galileo. Galileo had a
tremendous interest in what could be learned from experience. In addition
to his many well-known contributions to mathematics and physics, he also
reached "a correct solution to a gaming problem by laborious tabulation
of possible cases, but did not proceed to general principles."[100] Suppose
now that, instead of a gaming problem, a farming problem was given to
Galileo, by an apple farmer in Northern Italy.

Part of the annual life-cycle of an apple tree includes what in some
places is called the "June Drop." Early in a growth season, many young
apples fall to the ground long before reaching maturity. There are many
reasons for why this happens, including, for example: pollination may not
have occurred properly, pests, and malformed fruit. In fact, June Drop is
normal, and apple trees have several periods when fruit drops before a

---

[96] For general comments, see Preface. For a few examples, witness the spread of searchings
in, John D. Barrow et al, *Science and Ultimate Reality: Quantum Theory, Cosmology, and
Complexity*. Cambridge: Cambridge University Press, 2004.

[97] E. T. Bell, *The Development of Mathematics* (New York: McGraw-Hill Book Co., 1945),
155.

[98] Pierre de Fermat (1601/7 – 1665).

[99] Blaise Pascal (1623-1662).

[100] E. T. Bell, 154-155. The catalyst for the eventual discovery of the mathematical theory
of probability was a gambling problem proposed to Pascal by Antoine Gombaud and
Chevalier de Méré. See, E. T. Bell, 155.

harvest season.[101] Suppose, however, that one year a farmer in Italy notices that his trees seem to be dropping fruit in greater numbers than usual. Concerned for his orchard, he consults with the already famous Galileo.

Let us suppose that Galileo followed up on the farmer's concern by wondering, "Quante mele cadere da un albero?",[102] "How many apples fall from a tree?" And then, with added focus: "Quante mele cadono ogni giorno?", "How many apples fall each day?" For simplicity, let us imagine just one tree. Having a disposition that allowed Galileo to laboriously work out a solution to a gaming problem[103] (and also to patiently repeat experiments for understanding free-fall), we can imagine him working patiently over several days, obtaining a table of results somewhat similar to tables of results he had obtained for his free-fall experiments[104]:

Table 1.2. Statistical results Galileo might have obtained. Numbers of apples fallen from a tree.

Day	Number of Apples Fallen from a Tree
Day 1	30
Day 2	40
Day 3	36
Day 4	29
Day 5	33
Day 6	41
Day 7	34
Day 8	36

With Galileo, one may already have an understanding of how apples fall, that is, distance fallen is proportional to the square of time. Here,

---

[101] Carol Savonen, "Tree fruit drop is nature's way of reducing a heavy fruit load," http://extension.oregonstate.edu/gardening/node/889, June, 2007. Explanatory results are known in botany. See, for example, Alessandro Abruzzese, Ilaria Mignani and Sergio M. Cocucci, "Nutritional Status in Apples and June Drop," *J. Amer. Soc. Hort. Sci.*, 120 (1), (1995):71-74.

[102] No doubt, the Italian grammar needs a scholar's hand.

[103] See note 100.

[104] See note 80.

though, within the June Drop story, is there not a different question? The question is not about the motion of an apple as it falls to the ground; nor is it even about why[105] young apples fall from the limb of a parent tree.[106] There is a different question that can emerge, and an insight to be had. Is there not? In fact, is not a core question 'Is?'

Already having some understanding of free-fall, one may ask: Is there free-fall? And, if so, how often is there free-fall? From the table of results, one may grasp that the numbers of apples that fell varied around a central number, approximately 35. In this way, Galileo might have broken through to the beginnings of statistical method. To see this, notice how we get the number '35.' It is not in reference to any particular frequency in the aggregate of frequencies. It is, instead, discovered in the aggregate. In some contexts, one might even call the number '35' an "ideal frequency."

The experience gained in the present section already helps reveal a core aspect of statistical understanding. In statistical understanding, an emphasis is not 'What is it?', but 'Is it?' And, in the story, when Galileo reports back to the farmer, the report is not just a number '35,' but '35 apples that fell, ..., more or less.' In other words, whether actual or ideal, the significance of the frequencies partly depends on answers to questions of the form 'What is it?' And, we have, then, a fruitful glimpse into a normative complementarity of classical and statistical science.[107] For, one

---

[105] For examples of recent work on this question, see note 101.

[106] See, for example, note 101.

[107] Throughout scientific and philosophic traditions so far, *empirical probability* (*CWL3*, ch. 2, sec. 4) regularly is confused with *probability of judgment* (*CWL3*, "Probable Judgments," sec. 10.6). See, for example, Claus Beisbart and Stephan Hartmann, eds., in "Introduction" to, *Probabilities in Physics* (Oxford: Oxford University Press, 2011), 4. Lacking a control of meaning and attempting to apply empirical probability to individual events and occurrences has led to ongoing confusion. For one example, there is Hugh Everett's widely influential but non-verifiable many-worlds interpretation of probability functions in quantum physics (Hugh Everett, *Theory of the Universal Wavefunction*, Princeton, NJ: Thesis, Princeton University, 1956 and 1973; Hugh Everett, "Relative State Formulation of Quantum Mechanics," *Reviews of Modern Physics* 29 (1957): 454–462. There are also ongoing philosophic debates about non-locality of Aharonov-Bohm-type effects (Y. Aharonov and D. Bohm, "Significance of the Electromagnetic Potentials in Quantum Theory" (*The Physical Review*, vol. 115, No. 3, August 1 (1959): 485-491. There is a vast contemporary literature on similar effects. For a more recent discussion see, for

may wonder what an apple tree is; what apples are; and what free-fall is. But, one also may wonder about how many apple trees there are in an orchard; how many apples there are; and how often apples did in fact fall. And, in statistical data, one may even discover numerical benchmarks for how many apples have been falling, more or less.

1.9 Glimpsings of our Journey

In Chapter 5 of *Insight*, Lonergan provides two theorems, one on "the abstract formulation of the intelligibility immanent in Space and in Time"[108];

---

example, Richard Healey, "The Aharonov-Bohm effect," ch. 2 in, Richard Healey, *Gauging What's Real: The Conceptual Foundations of Contemporary Gauge Theories*. Oxford: Oxford University Press, 2007). The wonders of quantum events are not in doubt. But, in order to make progress, these problems are in desperate need of control of meaning, attainable through self-attention. It eventually becomes (self-) evident that asking whether or not event A will, or will not, occur is answerable by a Yes, or a No of judgment, or some qualified Yes or No, such as 'probably Yes,' or 'probably No,' all of which emerge from Is?-questions. In other words, in such cases, the name 'probability' is an adjective for a quality of judgment. In the same way, that is, through self-attention, one can make the preliminary observation that one also seeks answers to What?-questions. "(T)he central moment is an insight. By that insight the inquirer abstracts from the randomness in frequencies to discover regularities that are expressed in constant proper fractions named probabilities" (*CWL3*, 82). See also note 40. As mentioned in the Preface, there is the usual difficulty of quoting from *Insight*. To get a fuller sense of this from the present context, note that a Table of Contents of a typical graduate text in statistical science lays out a climb to be made through chapters and exercises. Section 2.4 of *Insight*, "Statistical Heuristic Structures," refers not to a book but to statistical science, including "quantum statistics (Bose-Einstein, Fermi-Dirac)" (*CWL3*, 91). Climbing to a control of meaning in empirical probability, at the level of the times, eventually will require that we work with examples from up-to-date applications of statistical science; and that we also engage in the further challenge of corresponding exercises in self-attention in one's development in statistical science. Beginnings are to be made, however, first through elementary examples, such as given in Section 8. Similar comments apply to Chapter 4 of *Insight*, "The Complementarity of Classical and Statistical Investigations." Helpful in these matters, but going beyond introductory discussion, is: Philip McShane, *Randomness, Statistics and Emergence* (Notre Dame, IN: University of Notre Dame Press, 1970), ch. 7, "Verification in Statistics and the Two Concepts of Probability."
[108] *Insight*, 174.

and a second theorem on the "concrete intelligibility of Space ... and ...the concrete intelligibility of Time." Because these are advanced results, in the present book they will need to be discussed later, in Chapter 5. Intermediate chapters of the present book will, among other things, invite readers to explore of some of the background that will be needed eventually for taking up those two theorems within a control of meaning at the level of the times.

However, as segue to the next chapters, what might we glimpse now about our journey? Continuing within elementary contexts (for example, attending to one's understanding of Galileo's 16th century law of falling bodies; and working with basic counting problems), it will be possible to make progress in one's positioning. Through elementary exercises, the reality of "reversal"[109] can become (self-) evident. One may grow in precision in the reality that there is an "undividedness of the sensing and the sensed"[110]; and that the real is neither "the already out there now"[111] nor a displacement of this to an "already in here now."[112] Through ongoing shifts, it can become self-evident that there is a "reversal of roles in which the sensible container becomes the intellectually contained."[113] All the while, one will be making progress in being luminous in one's 'What-questions' and one's 'Is-questions,' and our whole dynamics that are our 'What to do?' questions.[114]

Progress would appear, then, to require creative re-readings of *Insight*.[115] A needed beginning remains the first paragraph of *Insight*. Sweeping paradigm shifts are not ingested by sweeping popularizations.[116]

---

[109] See notes 53, 69, 70, and 110.

[110] Gasquet, p. 3. See note 52.

[111] See note 47.

[112] *CWL3*, 523.

[113] *CWL3*, 537.

[114] See note 40.

[115] Creative re-reading of *Insight* eventually will be in the light of Lonergan's later doctrine, functional specialization. See Epilogue.

[116] See Herbert Butterfield, "Ideas of Progress and Ideas of Evolution," ch. 12 in: *The Origins of Modern Science, 1300-1800* (New York: MacMillan, 1959). An online version is available at https://archive.org/details/originsofmoderns007291mbp. In the last chapter, Butterfield acknowledges the great sweeper, Bernard le Bovier de Fontenelle. The tradition continues in much of philosophy and history of science.

I think here, illustratively, of Richard Feynman's three pedagogical volumes on physics.[117] Only the third volume faces seriously the problem of pedagogy, and this only because the paradigm shift was altogether obscure to Feynman, to his students, to his contemporaries. For emergence of the new empirical method, there will need to be analogous struggles about past and present scientific and philosophic practice.

Note also that the works of Archimedes, Galileo, Kepler, Newton, Laplace, Maxwell, Einstein, Weinberg, Salam and Glashow (to name a few physicists, for example) are part of historical developments in the sciences, technologies and philosophy. Their works are implicit in, and integral to, meanings reached and results obtained in present times. In the new method, then, a control of meaning at the level of the times will involve a reaching control of meaning in historical understanding, a luminous living memory of our past and present state. Conveniently, we may look to partial heuristics of such that were mapped out by Lonergan:

The history of any particular discipline is in fact the history of its development. But this development, which would be the theme of a history, is not something simple and straightforward but something which occurred in a long series of various steps, errors, detours, and corrections. Now, as one studies this movement he learns about this developmental process and so now possesses within himself an instance of that development which took place perhaps over several centuries. This can happen only if the person understands both his subject and the way he learned about it. Only then will he understand which elements in the historical developmental process had to be understood before the others, which ones made for progress in understanding and which held it back, which elements really belong to the particular science and which do not, and which elements contain errors. Only then will he be able to tell at what point in the history of his subject there emerged new visions of the whole and when the first true system occurred, and when the transition took place from an earlier to a later systematic ordering; which systematization was simply an expansion of the former and which was radically new; what progressive transformation the whole subject underwent; how everything that was explained by the old systematization is now explained by the new one, along with many other things that the old one did not explain - the advances in physics, for example, by Einstein and Max Planck. Then and then alone will he be able to understand what factors favored progress, what hindered it, and why, and so forth.

---

[117] Richard Feynman, *The Feynman Lectures on Physics*, edited by R.B. Leighton and M. Sands (San Francisco: Addison Wesley, 1964), in three volumes. Now freely available online, through Caltech: http://feynmanlectures.caltech.edu/.

Clearly, therefore, the historian of any discipline has to have a thorough knowledge and understanding of the whole subject. And it is not enough that he understand it any way at all, but he must have a systematic understanding of it. For that precept, when applied to history, means that successive systems which have progressively developed over a period of time have to be understood. This systematic understanding of a development ought to make use of an analogy with the development that takes place in the mind of the investigator who learns about the subject, and this interior development within the mind of the investigator ought to parallel the historical process by which the science itself developed.[118]

Even Aristotle will not escape[119] the lift of philosophy into a new control of meaning. For those familiar with Archimedes' brilliant but axiomatically presented hydrostatics,[120] it is evident that one must also rescue Archimedes from himself. The community growing toward generalized empirical method will find themselves not once, but repeatedly, back on the first page of the first chapter of *Insight*, in hot water with Archimedes.

---

[118] Bernard Lonergan, *Early Works on Theological Method 2,* vol. 23 in, *Collected Works of Bernard Lonergan* (Toronto: University of Toronto Press, 2013), 175- 177. The quote may also be found in Michael G. Shield's earlier translation, *Understanding and Method*, 1990, 130-2. The original Latin text of *De Intellectu et Methodo*,1959, has the material on page 55. See *CWL23*, note 1, p. 3.

[119] Place his *Posterior Analytics* in the context of, Bernard Lonergan, "The Form of Inference", in *Collection,* vol. 4, *Collected Works of Bernard Lonergan,* eds. Frederick E. Crowe and Robert M. Doran (Toronto: University of Toronto Press, 1988), ch. 1, 3 – 15.

[120] T. L Heath, *The Works of Archimedes* (London: C. J. Clay and Sons, Cambridge University Press, 1897), "On Floating Bodies, Book I," 253-262; "On Floating Bodies Book II," 263-300. Online, see: https://archive.org/details/worksofarchimede029517mbp.

Chapter 2

A Foundations Lift from the Adult
Columbidae

Abstract: The focus of this chapter is animals, with special attention given to the *pigeon*, or *dove*. As can be witnessed in section three, while there are continuing developments in philosophy of biology, there are also ongoing confusions. This is not merely a problem in philosophy. For, these confusions influence front-line field work in avian science, and biology generally. Sections three to eight focus on the biology of an *adult* pigeon. Part of this work takes us into the Kreb's cycle, or TCA cycle, verified in experiments with fresh wing-muscle tissue in phosphate saline solution. The middle sections of this chapter are partly motivated by the dense heuristics of *Insight* 489-90: "Study of an organism begins from the thing-for-us, from the organism as exhibited to our senses. ...." Included are preliminary forays in self-attention. The remoteness of Lonergan's heuristics becomes increasingly evident, but, at the same time, so does their relevance to progress in biology and philosophy of biology. Section nine of the chapter edges up to the main topic of the next chapter 3, organic development. For, while it is convenient to temporarily restrict attention to the adult pigeon, before reaching air mastery, the pigeon first is a blind embryo, grows into an awkward squab, then a struggling fledgling on the edge of flight.

2.1 Preliminaries

Pigeons are found throughout the world, in many shapes, sizes and colors. Their habitats include rock crevices, forests, villages, towns, and the concrete jungles that we call cities. For many, seeing pigeons in flight is a familiar experience. And, yet, since the earliest of times, bird flight has captured human imagination,[1] as it may the reader's.

As discussed in chapter one, Galileo (1564-1642) made a beginning toward understanding how apples fall to the ground. Later, Newton (1642-1727) explained free-fall within a system, with a law of universal gravitation. In Newton's theory, the motion of an apple that falls to the ground is explained in the same way as the moon that falls toward the earth, and planets that fall toward the sun. More recently, there was Einstein's well-known breakthrough to a covariant theory of gravitation. What, though, can be said about a rather familiar feathery animal that not only does not fall toward the ground, but dives toward the ground; and that can as swiftly turn and lift to new aerial maneuvers of, not free-fall, but free-flight? For the ornithologist, a family name for the animal is *columbidae*, which comes from the Greek for "dive."[2] The English name *pigeon* is from Latin, for "peeping," the sound chicks make. The other common name for the bird is *dove*, a Germanic word that, like the Greek word, also means "dive."

This chapter continues our foray toward generalized empirical method. The focus shifts from questions about space and time, to questions about animals that we call *pigeon*, or *dove*. Among their many abilities, true to their family name *columbidae*, these animals can be seen to have an astonishing mastery of space and time.

There is, of course, a long tradition of learning and writing about birds.[3] The tradition includes Aristotle's work, where approximately 25 percent of the extant corpus contains his results in zoology. His detailed investigations of animals, include, in particular, precise descriptions of the eyes of a four day old chick embryo. Also part of his work in zoology was his searching for a set of standards by which to study living things, an investigation of modes of causality and of necessity proper to biological explanation, the relation of form to matter in living things, the proper

---

[1] For some details, going back to prehistoric times, see, John J. Videler, *Avian Flight* (Oxford Ornithology Series). Oxford: Oxford University Press, 2005.

[2] Greek: *kolumbos*, "a diver."

[3] See note 1.

division of the subject matter, the means of identifying kinds and their activities at the proper level of abstraction, and much more.[4]

In contemporary avian science, there are ongoing advances within hundreds of sub-disciplines[5] (the number is increasing) in, for example, avian biophysics; avian biochemistry; avian neuroscience; birds and ecology; the evolution of birds; avian veterinary science; history and philosophy of avian science, philosophy of biology generally. To complexify matters further, within a basic position,[6] one grows in the reality that all experience provides data within human consciousness. And generalized empirical method asks that the scientist appeal to all data.[7] That sounds simple enough, and it would be difficult to deny the value of such a precept. But, which data? In avian science, there are increasingly many sources of data obtained in doing biophysics, biochemistry, avian population studies, ecology, avian evolution, philosophy of biology, and so on. A further challenge for control of meaning comes from the fact progress in biology is historical. And, as noted in Chapter 1, the

> history of any particular discipline is in fact the history of its development. ... (S)ystematic understanding of a development ought to make use of an analogy with the development that takes place in the mind of the investigator who learns about the subject, and this interior development within the mind of the investigator ought to parallel the historical process by which the science itself developed.[8]

---

[4] See, for example, James Lennox, "Aristotle's Biology," *The Stanford Encyclopedia of Philosophy* (Spring 2014 Edition), Edward N. Zalta (ed.), http://plato.stanford.edu/entries/aristotle-biology/. See also James Lennox, trans., *Aristotle: On the Parts of Animals I-IV*. Gloucestershire: Clarendon Press, 2002. Aristotle's results in zoology are, or course, dated. However, aspects of his work about potency, form and act remain significant. See, for example, sec. 4.4, below.

[5] As mentioned in the Preface, the possibility of effective collaboration through functional specialization was identified by Lonergan in 1965. See Epilogue.

[6] *CWL3*, 413.

[7] Bernard Lonergan, *A Third Collection*, end of 140.

[8] Bernard Lonergan, *Early Works on Theological Method 2, Collected Works of Bernard Lonergan*, Vol. 23 (Toronto: University of Toronto Press, 2013), 175-177. See also Michael G. Shield's earlier translation, *Understanding and Method*, 1990, 130-2. The original Latin text of *De Intellectu et Methodo* (1959) has the material on page 55. See *CWL23*, note 1, p. 3. The entire quotation is also in the present book, in Section 1.8.

Note also that, as can be witnessed in section 2.2, below, while there is development in contemporary philosophy of biology, there are also various ongoing confusions. And, that confusion is not merely a philosophic problem. Among other things, philosophy of biology influences front-line field work and even general education.[9]

An introductory book is not the place to attempt to resolve fundamental issues in contemporary philosophy of biology. To do so effectively is beyond the reach of any single work: biology is a global historical collaborative enterprise.[10] Moreover, this book is only an invitation to a *future* method. So, while it would be helpful, it is not assumed that the reader already has expertise in 20th and early 21st century biology. Even so, it is possible to at least get some impression of various confusions in present day philosophy of biology. This is done in the next section, by looking to a sampling of texts from the contemporary literature.[11] Reading with self-attention, one can begin to witness the fallacy of certain trends, including the fiction of derived notions such as "cellular mechanism," "information networks," "genetic code," "epigenetic code," and the like. Introductory forays can help increasingly reveal both the pressing need and the possibility of control of meaning in philosophy of biology.

2.2   Some Views in Philosophy of Biology

2.2.1   *On views in philosophy of biology*

The literature includes a wide range of diverse views. To get a sense of ongoing controversies, one may look, for example, to *Contemporary*

---

[9] See Section 5.7 and Epilogue.

[10] See notes 5 and 9.

[11] Note that the purpose here is not interpretation. Interpreting the work of a scholar is its own task. Within a future generalized empirical method, some of the great demands of that task are indicated in *CWL3*, sec. 17.3. Effective interpretation will be under functional control, some aspects of which are described briefly, in Bernard Lonergan, *Method in Theology* (Toronto: University of Toronto Press, 2003), ch. 7, "Interpretation" (first pub.: London: Darton, Longman and Todd, 1972)). See Epilogue. Here, I use samplings of various authors' works more simply, to offer a few points of entry into preliminary (foundational) exercises.

Debates in Philosophy of Biology.[12] The book is a collection of pairs of papers representing what are taken to be more or less mutually opposing views.[13] There is, for instance, the paper, "It Is Possible to Reduce Biological Explanations to Explanations in Chemistry and/or Physics,"[14] immediately followed by the paper, "It Is Not Possible to Reduce Biological Explanations to Explanations in Chemistry and/or Physics."[15]

The abstract for "It is Possible ..." includes the following:

> Accordingly, we need an account of the evolution of (biological) function out of simple physical and chemical dynamics which we do not as yet have. Nevertheless, I retain a guarded optimism: we are, I believe, moving closer to answering this basic question. Yet doing so, I argue, requires fundamental transformations in the conventional approaches of both physics and chemistry.[16]

Near the beginning of the opposing view, "It is not possible ...," Dupré writes:

> Like Keller, I am a materialist. That is to say, I do not believe there is any kind of stuff in the world other than the stuff described by physics and chemistry. There are no immaterial minds, vital forces, or extra-temporal deities. Keller also writes, however, that as a materialist she is "committed to the position that all biological phenomena, including evolution, require nothing more than the workings of physics and chemistry." Even as a materialist, I'm not sure I feel committed to this; but, of

---

[12] Francisco J. Ayala and Robert Arp (Editors), *Contemporary Debates in Philosophy of Biology*. Hoboken, NJ: Wiley-Blackwell, 2009.

[13] While there are many views, there are methodological commonalities that contribute to ongoing confusions. Some aspects of the problem will begin to emerge in paragraphs below.

[14] Evelyn Fox Keller, "It Is Possible to Reduce Biological Explanations to Explanations in Chemistry and/or Physics," ch. 1 in, Francisco J. Ayala and Robert Arp (Editors), *Contemporary Debates in Philosophy of Biology* (Hoboken, NJ: Wiley-Blackwell, 2009), 19-31.

[15] John Dupré, "It Is Not Possible to Reduce Biological Explanations to Explanations in Chemistry and/or Physics," ch. 2 in, Francisco J. Ayala and Robert Arp (Editors), *Contemporary Debates in Philosophy of Biology* (Hoboken, NJ: Wiley-Blackwell, 2009), 32-48.

[16] Evelyn Fox Keller, 19.

course, that depends upon exactly what the title question means. A little unpacking of this question may help to reveal where (if anywhere) there is a serious difference between Keller's position and my own.[17]

Introducing the third pair of papers, "Part III, Are Species Real?," the editors of the book write the following:

It may be that the definition and nature of species will always be a problem for researchers because of the "subjective dependence" of species concepts on the particular view of the taxonomist, as noted by Claridge. Yet, given the myriad attempts at a clear and coherent universalizable definition of species still occurring, it may be that researchers are resistant to this subjectivity.[18]

The two papers of Part III are: "Species Are Real Biological Entities,"[19] and "Species Are Not Uniquely Real Biological Entities."[20] The Conclusions section of the "Species Are Real" paper begins as follows:

The species problem has always confused two almost completely separate phenomena— species concepts and species taxa— as emphasized frequently over a period of more than 60 years by Ernst Mayr. Species taxa are recognized and described by taxonomists according to their own preferred species concepts. Taxonomists may also be influenced in their choice of concept by the particular groups of organisms on which they work. I follow Mayr in the view that many species concepts are in reality recipes for recognizing particular species taxa and not themselves significant and distinct concepts.[21]

In the last paragraph of the same paper, we find:

We need broadly applicable species concepts and the existing Linnean system of nomenclature, certainly for species names.[22]

---

[17] John Dupré, 33.

[18] Francisco J. Ayala and Robert Arp (eds.), Editors' introduction to Part III, in, *Contemporary Debates in Philosophy of Biology* (Hoboken, NJ: Wiley-Blackwell, 2009), 88.

[19] Michael F. Claridge, ch. 5 in, *Contemporary Debates in Philosophy of Biology, 91-109.*

[20] Brent D. Mishler, ch. 6, in *Contemporary Debates in Philosophy of Biology*, 110-122.

[21] Claridge, 102.

[22] Claridge, 105.

In the next paper, the author, Mishler,[23] takes another view, and argues that "Species Are Not Uniquely Real Biological Entities."

> The so-called "species problem" is really just a special case of the taxon problem. Once a decision is made about what taxa in general are to represent, then those groups currently known as species are simply the least inclusive taxa of that type.[24]

These have been only a few samples taken from one essay collection. The diversity there, however, is representative. And, indeed, further sampling of the philosophy of biology literature reveals increasing ranges of admissible variations in views about what species might be. This problem also is alluded to in Claridge's article:

> No topic in evolutionary and systematic biology has been more contentious and controversial than the nature and meaning of species. ... In addition to arguments about different philosophical approaches, much of the controversy has centered on the confusion between, on the one hand, the philosophical concepts of species and, on the other, the practical recognition of species taxa themselves. The frequent confusion of these two different aspects of the species problem continues to cause much argument and controversy among biologists and philosophers.[25]

How, though, do biology and the philosophy of biology relate? Peter Godfrey-Smith, a contemporary philosopher of biology, writes:

> In some of the areas described ... the goal of the philosopher is to understand something about *science*[26] – how a particular part of science works. In other cases, the goal is to understand something about the natural world itself, the world that science is studying.[27]

Earlier in the same chapter he writes:

---

[23] Brent D. Mishler, "Species Are Not Uniquely Real Biological Entities," ch. 6 in, Francisco J. Ayala and Robert Arp (Editors), *Contemporary Debates in Philosophy of Biology* (Hoboken, NJ: Wiley-Blackwell, 2009), 110-122.

[24] Mishler, 110.

[25] Claridge, 92.

[26] Italics in source text.

[27] Peter Godfrey-Smith, *Philosophy of Biology* (Princeton and Oxford: Princeton University Press, 2014), 3. I draw attention to some of the work of Godfrey-Smith, not to single out his work, but because his methods are representative of common trends.

philosophy is concerned with 'how things in the broadest possible sense of the term hang together in the broadest possible sense of the term.' Philosophy aims at an overall picture of what the world is like and how we fit into it.[28]

After discussing a number of views and related issues in the field, in the last chapter of his book[29] Godfrey-Smith offers further details about his own view:

> I'll argue in this final chapter against some of the most strongly information-infused views of biology. Then, however, I'll look at the unifying role of a related idea: communication.[30]

Referring to the results of some recent neurobiologists and psychologists, he writes:

> the message of recent brain science is … that brains have devised a different way of solving the problem of storing memories, and the write-read model does not apply (Koch 1999). This second view might be expressed by saying that rather than a *write-read* mechanism the brain uses a *write-activate* memory. The marks left by experience in memory can do their job without a reader. If this is right, the brain is a kind of flipside to the case of genetic systems, as it is the reader that has been avoided; *write-activate* rather than *evolve-read*.
>
> With this comparison laid out it is interesting to add another memory system within cells, a system more readily seen *as* memory: the modification of DNA with chemical marks (especially methyl groups) that inhibit transcription. Some people call this an "epigenetic code." In that case, the "writing" step is clear; the DNA is marked in a systematic way by machinery with that function. …
>
> …
>
> So within the varieties of memory there are systems that have a write-read character (which I see as fitting the sender-receiver model), and there are variants that get by without a write step or a read step, perhaps both.
>
> …
>
> A basic feature of life in multicellular animals is differentiation. … This works by

---

[28] Godfrey-Smith, *Philosophy of Biology*, 1.

[29] Godfrey-Smith, *Philosophy of Biology*.

[30] Godfrey-Smith, *Philosophy of Biology*, 144.

the regulation of gene expression; …. This involves memory, as seen above, and it involves signaling over space as well.

…

There are continuities between the simple kinds of signaling seen between cells, through animal communication, to the highly elaborated forms seen in human communication – gesture, speaking, picturing, writing – that arise from the special forms of social involvement characteristic of our species. … Communication-like behaviors are ubiquitous, and communication is also a manifestation of something more basic. A combination of receptivity and activity, with those behaviors stabilized by selection, by feedback, is a distinctive feature of the living world.[31]

In an earlier article, Godfrey-Smith gives some details on what he meant by "genetic system" and "signaling":

Drawing on models of communication due to Lewis and Skyrms, I contrast sender-receiver systems as they appear within and between organisms, and as they function in the bridging of *space*[32] and *time*.[33] Within the organism, memory can be seen as the sending of messages over time, communication between states as opposed to spatial parts. Psychological and genetic memory are compared with respect to their relations to a sender-receiver mode. Some puzzles about "genetic information" can be resolved by seeing the genome as a cell-level memory with no sender.[34]

In the concluding paragraphs of the article, he writes:

The best starting point for a view of the gene action that takes concepts of coding and representation seriously, as acknowledged in much of the earlier literature, is the reality of the "reading" step. Genes have a rather clear reader mechanism, the transcriptional and translational machinery. The next move is not so clear. My approach is to make an explicit link to the idea of memory. This link was made very early in the discussion, by David Nanney. He noted in 1958 that there are two distinct tasks that cells manage to perform:

---

[31] *Philosophy of Biology*, 155-156.

[32] Italics in source document.

[33] Italics in source document.

[34] Peter Godfrey-Smith, "Sender-Receiver Systems Within and Between Organisms," in *Philosophy of Science Assoc. 23rd Biennial Meeting.* San Diego, CA: Philosophy of Science Association, 2012. http://philsci-archive.pitt.edu/view/confandvol/confandvol 2012psa23rdbmsandcalpsasymposia.html.

> On the one hand, the maintenance of a "library of specificities," both expressed and unexpressed, is accomplished by a template replicating mechanism. On the other hand, auxiliary mechanisms with different principles of operation are involved in determining which specificities are to be expressed in any particular cell. ... [These] will be referred to as "genetic systems" and "epigenetic systems" (1958, 712).
>
> This is a helpful way of looking at the situation, now as well as then. A genome can be seen as containing a memory of the structure of useful protein molecules. ... Specifically, ... , a genome is a *cell*-level memory, not a memory for the whole multicellular organism or population. ... Organisms are the results, the upshots, of what the cells are doing.
>
> So far, this seems to fit well with a temporally organized version of a sender-receive structure. ... Who, then, is the sender? ... The answer is that there is no sender.[35]

A challenge before us is to make a beginning in reading these selections, with self-attention. Within such an effort, various questions arise. What, for example, are grounds for asserting that "we need an account of the evolution of (biological) function out of simple physical and chemical dynamics?"[36] If that is one's goal, what might it mean to say "simple"[37] physical and chemical dynamics, given the complexities and unsolved problems of contemporary quantum field theories, quantum-physical-chemistry, and chemical dynamics of macromolecules?

There is little doubt that physics and chemistry are essential to all material entities. But, looking to gardens, fields, forests, waterways and skyways of the world, and the astounding diversity of living things in the world, is it plausible that all of this is only physics and chemistry? If, in physics say, one reaches some up-to-date explanation of gravitational *free-fall*, is one then also explaining gravity-defying *free-flight*? In biochemistry, one may attain some understanding of the Krebs cycle.[38] One of the (simplified) equations in the cycle is:

---

[35] Peter Godfrey-Smith, "Sender-Receiver Systems Within and Between Organisms."
[36] Note 16.
[37] Notes 16 and 31.
[38] The Krebs cycle will be discussed in some detail, in Section 2.5.

$$\text{fumarate} + \text{pyruvate} + 2O_2 \rightarrow \text{succinate} + 3CO_2 + H_2O.$$

But, by which understanding in chemistry does one explain how it is that a bird acquires needed concentrations of reactant O_2? Among other things, note, and self-note, that chemical equations presuppose boundary conditions.[39]

Many other questions arise. For instance, is it self-coherent to *argue* that there are only physics and chemistry? As can be descriptively verified by attending to one's own experience in doing physics and chemistry, neither physical nor chemical understanding explicitly regard the capacity to *argue*. On what grounds can one insist *a priori* that there needs to be a "universalizable definition of species"? Are the "species problem," "species concepts" and "taxa" to be sorted out by *decisions*[40] about terminology? No doubt, no one person understands everything in a science, making scientific belief essential to progress. But, within science and the philosophy of science, is it coherent with scientific practice in physics, chemistry and biology to resort to *belief*, such as in:

> I do not believe there is any kind of stuff in the world other than the stuff described by physics and chemistry?[41]

In the work of Godfrey-Smith, what might it be like to know how things work "in the broadest possible sense of the term?" In the last chapter of his book, Godfrey-Smith writes of "the unifying role of a related idea: communication."[42] But, what about a real organism? On what grounds can one assert that: "(t)he best starting point for a view of the gene action that takes concepts of coding and representation seriously, as acknowledged in much of the earlier literature, is the reality of the reading step;"[43] and then state, as factual, that genes "have a rather clear reader mechanism, the transcriptional and translational machinery"?[44] Godfrey-Smith goes on to

---

[39] All of these questions hint at the need of layerings of series of exercises in self-attention. Beginnings toward such preliminary exercises are a main content of sections 2.3 to 2.8.

[40] See note 24.

[41] See note 17.

[42] See note 30.

[43] See note 35.

[44] See note 35.

say that the "next move is not so clear. My approach is to make an explicit link to the idea of memory." He follows with the assertion that there is "a *write-read* mechanism, a template replicating mechanism, signaling over *space* and *time*."[45]

Are there data of some kind to support these various claims and choices? Certainly, progress is being made in the philosophy of biology. But, what are the heuristics? Within main debates in the tradition, there is appeal to concepts,[46] diagrams[47] and language usage. Might it not also be important for a philosophy of biology to appeal to how results are obtained in biology? What are examples of biological data? Which questions are biological questions? Which data are diagrams? Which data are experimental? Which data are symbolic or linguistic? What are the insights? And so on. As we track through and self-track through readings such as those referenced above, do not key features of (what at the moment are) standard philosophical arguments begin to show as being rather distant from actual biological understanding of actual biological organisms?

But, let's look at some of this in a little more detail. Philosophical arguments about cellular mechanisms and signaling networks appeal to diagrams of networks of chemical reaction equations.[48] Such diagrams (often called models[49]) are said to represent networks of chemical

---

[45] Italics in source document.

[46] Quentin D. Wheeler and Rudolph Meier, eds., *Species Concepts and Phylogenetic Theory: A Debate*. New York: Columbia University Press, 2000. Through debate, the book presents five well-known definitions of the "species concept" (Ernst Mayr; Rudolph Meier and Rainer Wilmann; Brent Misher and Edward Theriot; Quentin Wheeler and Norman Platnick; E. O Wiley and Richard Mayden).

[47] See, for example, note 48.

[48] See, for example, the TCA cycle, note 86.

[49] I merely cite usage of the name 'model' in the present context. In this book, I do not raise questions about what models are in the sciences and the philosophy of science. A reference for 'models in science,' with an extensive bibliography can be found in Roman Frigg and Stephan Hartmann, "Models in Science," *The Stanford Encyclopedia of Philosophy* (Fall 2012 Edition), Edward N. Zalta (ed.), http://plato.stanford.edu/archives/fall2012/entries/models-science/. See also Daniela Bailer-Jones, *Scientific Models in Philosophy of Science*. Pittsburgh, PA: University of Pittsburgh Press, Digital Research Library, 2012.

reactions in cells and multicellular organisms. But, in living organisms, there are no experimental data on either cellular mechanisms as imagined, or signaling networks[50] as imagined. Note also that biochemistry has been moving forward despite the fact that

> current technologies are in general unable to interrogate (the biochemistry of) individual cells.[51]

Of course, this is not to suggest that there is not an abundance of experimental data in the established science of biochemistry. However, even if technological developments provide images of parts of individual cells, questions in biochemistry are not answered by appealing to images, but by verifying reaction equations. And, in organisms, such results are fragmentary (such as for components in the TCA cycle[52]). Component reaction equations are partly verified through sequences of groupings of reaction baths remote to a living organism. There is numerical data obtained through computer simulations. Such numerical data, however, is not data from living organisms, but from computer programs for hypothetical networks starting from supposed boundary conditions.

Raising questions here is not to suggest that there is not important work in the philosophy of biology.[53] But, evidently, there are core problems in method. For, in particular, present methods allow for ongoing debate about non-verifiable conceptual constructs and models. [54] Does not philosophy

---

[50] For another example, the major pathways of cellular respiration of a typical eukaryotic cell are given in Figure 2.3.

[51] "(C)urrent technologies are in general unable to interrogate (the biochemistry of) individual cells. Therefore, transcriptome characterizations emerging from microarray or RNASeq surveys are very coarse, providing information of the average behavior of ensembles of tens, hundreds of thousands or even millions of cells. ... in interrogating complex tissues, the behavior of the ensemble may be very different from that of individual cells" (Roderic Guigó, "The Coding and the Non-coding Transcriptome," ch. 2 in A. J. Marian Walhout, Marc Vidal and Job Dekker, eds., *Handbook of Systems Biology, Concepts and Insights* (Amsterdam: Academic Press, 2013), 38.)

[52] In sections 2.4 and ff, the TCA cycle will be explored in some detail.

[53] See note 65, and brief comments there regarding, for example, the contributions of Denis Noble (Department of Physiology, Anatomy and Genetics, Oxford), and his decades long searchings for *The Music of Life*.

[54] The problem is described by Ernst Mayr (1904-2005): "No matter in what branch of

of biology cry out for an improved control of meaning and some kind of development in empirical method? Whatever one may understand or decide about species concepts, logical models, diagrams or computational schemes, obviously, a living bird is none of these. In name-focused sight-held-feathered-wonder-lift, the question arises: "What is this thing called '*d-o-v-e* '?"[55]

---

biology one is interested, it is necessary to work with species. ... Considering this outstanding importance of the species in biology, it strikes me as almost scandalous that there is still so much disagreement and uncertainty about almost every aspect of species. There is no other problem in biology on which more has been written in recent years and less unity has been achieved than the species problem (Ernst Mayr, *What Makes Biology Unique? Considerations on the autonomy of a scientific discipline* (Cambridge: Cambridge University Press, 2004), 8). Later, in "Another look at the species problem" (Mayr 2004, ch. 10, 171-193), Mayr notes the following: "The species, together with the gene, the cell, the individual, and the local population, are the most important units in biology. Most research in evolutionary biology, ecology, behavioral biology, and almost any other branch of biology deals with species. How can one reach meaningful conclusions in this research if one does not know what a species is and, worse, when different authors talk about different phenomena but use for them the same word – species? But this, it seems, is happening all the time, and this is what is referred to as the species problem. There is perhaps no other problem in biology on which there is as much dissension as the species problem. Every year several papers, and even entire volumes, are published, attempting to deal with this problem. The species problem is indeed a fascinating challenge. In spite of the maturation of Darwinism, we are still far from having reached unanimity on the origin of new species, on their biological meaning, and on the delimitation of species taxa. The extent of the remaining confusion is glaringly illuminated by a recent book on the phylogenetic species concept (Wheeler and Meier, *Species Concepts and Phylogenetic Theory: A Debate*. 2000. Reference given in Mayr, 2004. See note 46). ... The result is great confusion" (Mayr, 2004, 171-172).

[55] The wording here is to intimate something of the dynamics of naming and reaching for understanding of what one has named. This touches on enormously challenging problems. Reaching explanatory heuristics for 'naming' is a remote achievement for the human sciences. The problem is touched on briefly in the discussion in Section 3.5 of Bernard Lonergan's book *Method in Theology*: "The moment of language in human development is most strikingly illustrated by the story of Helen Keller's discovery that successive touches made on her hand by her teacher conveyed names of objects" (Lonergan, *Method in Theology*, 70). The touches were for w-a-t-e-r. See, also *Insight*, *CWL3*, 578, where Lonergan draws attention to "experiential conjugates that involve a triple correlation of classified experiences, classified contents of experience, and corresponding names." Eventually, the problem is to be held within a heuristics of explanatory interpretation: "To

2.2.2 On philosophy of biology influencing field biology

Up to now, questions in this section have focused on views in philosophy of biology. Next, I would like to draw attention to work that helps illustrate how developments and problems in philosophy of biology are not merely philosophic issues. That is, philosophy of biology has been influencing what often is called biological research itself.[56] For some data on this, we may look to *systems biology*, a name for a loosely defined family of views[57] that emerged in the 20th century, and that now are increasingly influential in contemporary field-work.

In its most familiar form, contemporary systems biology came out of the philosophical work of Bertalanffy (1901-1972) on "general systems theory."[58] An early 20th century contributor to the development of systems

---

avoid confusion and misunderstanding, it will not be amiss to draw attention to the possibility of an explanatory interpretation of non-explanatory meaning" (*CWL3*, 610). A helpful introduction to the problem of naming can be found in: Philip McShane, *A Brief History of Tongue* (Axial Publishing, 1998), 31-37. The sentence in the text above also refers to 'thing.' In opaque brevity, one might say that a "notion of a thing is grounded in an insight that grasps, not relations between data, but a unity, identity, whole in data, ..., (taken) in their concrete individuality and in the totality of their aspects" (*CWL3*, 271). See also *CWL3*, 461, a major shift of context, discussed somewhat in Section 4.4, below. However, the present chapter continues in elementary mode, and, in particular, continues in "dodging the question. What is a thing?" (*CWL3*, 270). For, the aim here continues to be elementary description, gathering data that will help invite future development in a not-yet-implemented generalized empirical method. Regarding realities intended, see sections 4.4 and 4.7.

[56] See also Preface and Epilogue.

[57] A brief description of some of the recent internal debates within applied systems biology can be found in: Jane Calvert and Joan H. Fujimura, "Calculating life? Dueling discourses in interdisciplinary systems biology," *Studies in History and Philosophy of Biological and Biomedical Sciences 42* (2011): 155–163.

[58] See, e.g., his article, Ludwig von Bertalanffy, "General system theory - A new approach to unity of science," *Human Biology*, (Symposium), December, Vol. 23 (1951): 303-361. One of his well-known books is: *General Systems Theory* (New York: George Braziller, 1968). Note, however, that for various historical reasons, Denis Noble locates an early systems biology in the searchings of Claude Bernard, 1865. See, Denis Noble, "Claude Bernard, the first systems biologist, and the future of physiology," *Experimental Physiology*, 93 (Jan. 1, 2008): 16-26. http://ep.physoc.org/content/93/1/16.full.pdf+html.

theory in biology was Niclolas Rashevsky (1899-1972).[59] Among other things, Rashevsky was studying the physics of cell membrane motion. While his work focused on cell membrane motion, his approach remains prevalent throughout contemporary systems biology:

> Systems Biology ... places the theoretical foundations of systems analysis of living matter into the context of modern high-throughput quantitative experimental data, mathematics, and *in silico* simulations. The aim is to analyze the organization and to gain engineering-control of metabolic and genetic pathways.[60]

Hypotheses are made about supposed ideal aggregates. Then, mathematical and computerized computations yield ranges of possible mean paths and possible time-dependent ideal concentrations.

How, though, can such results be interpreted? In systems biology, usually it is assumed that such results represent individual events, imagined within individual cells.[61] But, if, for example, computations reveal limitations on possible averages of boundary conditions and biochemical reaction rates, by both definition and experimental verification, averages are significant only when they are verified in representative samples. And, in such applications, there are allowed margins of error. Any sample is itself, typically, some indeterminately large aggregate. In any case, neither theoretically nor experimentally do mathematically derived results about mean paths and time-dependent ideal concentrations refer to any single aggregate, let alone the cellular membrane of an individual cell within any single aggregate. Note also that even when, within allowed margins of error, results on averages are

---

[59] Rashevksy (Nicolas Rashevsky, 1899-1972). See, Nicolas Rashevsky, *Mathematical Biophysics: Physico-mathematical Foundations of Biology,* Vol. 1 of 2 volumes, 3rd rev. edition. New York: Dover Publications, 1960. Rashevsky supposes availability of various physical and chemical boundary conditions for averages (chemical reaction rates, chemical flux rates, mass densities, etc.).

[60] Melanie Boerries, Roland Eils and Hauke Busch, "Systems Biology," ch. 1 in, Robert A. Meyers, ed., *Systems Biology, Current Topics from the Encyclopedia of Molecular Cell Biology and Molecular Medicine* (Somerset, NJ: John Wiley & Sons, 2012), 5).

[61] For example: "When these ensembles are relatively uniform – as in cultures of immortalized cells – average behaviors may be confidently extrapolated to the behaviors of individual cells" (Roderic Guigó, "The Coding and the Non-coding Transcriptome," in *Handbook of Systems Biology,* p. 38.

verified in representative samples of cell populations, because of rapidly changing intra-cellular boundary conditions, ranges of possible coefficients of each such system of equations are only valid for approximating averages for extremely small time intervals.

On the other hand, it is known that such methods have led to important results in, for example, working out relatively precise limitations for various metabolic rates.[62] And, these methods are contributing to advances in, for example, biochemistry for medicine.[63] But, as the previous paragraph points to, it is evident (and self-evident) that results along these lines are extremely remote from any particular cell, and even more so from any particular multi-cellular organism.

Further data on the problem can be obtained by looking to the more recent *Handbook of Systems Biology*.[64] The book is a collection of articles

---

[62] In the last decade there have been considerable developments in systems biology. For example, there are now 'multi-scale systems.' In that case, a *systems* premise is the same, but reaction terms from physics are included. One of the computational challenges for such models is that physical events, chemical events, and biological events take place on different time scales. For instance, in protein folding, "the time scale for the vibration of the covalent bonds is on the order of femto-seconds (10^{-15} s), folding time for the proteins may very well be on the order of seconds." (E. Weinan and Bjorn Engquist, "Multiscale Modeling and Computation," *Notices of the Amer. Math. Soc.*, Vol. 50, No. 9, Oct. (2003): 1062-1069, [http://www.ams.org/notices/200309/fea-engquist.pdf]). Supercomputers can be used to approximate averages, through large numbers of iterated computations and estimates for averages, expectation values for energy levels, and so on.

[63] For example, systems biology continues to have "profound implications for cancer research" (Pau Creixell at al., "Navigating cancer network attractors for tumor-specific therapy." *Nature Biotechnology* 30, (2012): 842-848). For example, see K. A. Janes, et al. "A systems model of signaling identifies a molecular basis set for cytokine-induced apoptosis," *Science* 310, (2005):1646–1653. Similar results are ongoing. See, for example, Joseph Loscalzo and Albert-Laszlo Barabasi, "Systems Biology and the Future of Medicine," *Wiley's Interdisciplinary Reviews: Systems Biology and Medicine*, Vol. 3 (2011): 619-627. Another recent review article, on systems biology in biochemistry for medicine, is Miguel Angel Medina, "Systems biology for molecular life sciences and its impact in biomedicine," *Cellular and Molecular Life Sciences*, 70 (2013):1035–1053.

[63] Martha L. Bulyk and A. J. Marian Walhout, "Gene Regulatory Networks," ch. 4 in *Handbook of Systems Biology, Concepts and Insights*, eds. Marian Walhout, Marc Vidal and Job Dekker (Amsterdam: Academic Press, 2013), 65-88.

[64] A. J. Marian Walhout, Marc Vidal and Job Dekker (eds.), *Handbook of Systems Biology, Concepts and Insights*. Amsterdam: Academic Press, 2013.

written by "leading experts in the field who have contributed individual chapters that cover a range of topics and technologies."[65] Again, for present purposes, we only look to a small sampling.

Starting with the table of Contents of the *Handbook of Systems Biology*, one already sees that the terminology of networks and computer terminology dominate throughout.[66] Some of the words that recur in the collection of articles are 'circuit,' 'coding,' 'system,' 'network,' 'information flow,' 'graph theory,' 'Boolean network,' 'cell signaling,' 'computation,' 'network logic.' The first section of Chapter 4, by Bulyk

---

[65] Walhout, Vidal and Dekker (eds.), *Handbook of Systems Biology*, xi. The current literature on systems biology is large. A few texts representative are: Frank C. Boogerd, Frank J. Bruggeman, Jan-Hendryk S. Hofmeyr and Hans V. Westerhoff. *Systems Biology, Philosophical Foundations*. Amsterdam: Elsevier, 2007. Robert A. Meyers, ed., *Systems Biology, Current Topics from the Encyclopedia of Molecular Cell Biology and Molecular Medicine* (Somerset, NJ: John Wiley & Sons, 2012). Anirvan M. Sengupta, *Modeling Biomolecular Networks, An Introduction to Systems Biology*. Oxford: Oxford University Press, 2016. Isidore Rigoutsos, Gregory Stephanopoulos, eds., *Systems Biology*, vol. 1, *Genomics*. Oxford: Oxford University Press, 2006. Isidore Rigoutsos, Gregory Stephanopoulos, eds., *Systems Biology*, vol. II: *Networks, Models, and Applications*. Oxford: Oxford University Press, 2006. Glenn Rowe, *Theoretical Models in Biology, The Origin of Life, the Immune System, and the Brain*. Oxford: Clarendon Press, 1998. Of particular influence in the field has been the work of Denis Noble (Balliol College and Department of Physiology, Anatomy and Genetics, Emeritus Professor, University of Oxford). His many publications include, *The Logic of Life: The Challenge of Integrative Physiology* (Oxford: Oxford University Press, 1993); and, more recently, *The Music of Life: Biology Beyond the Genome* (Oxford: Oxford University Press, 2006). His work comes from a decades long ongoing searching. Partly by looking to contemporary experimental results, he helps bring out the need for a "truly multi-level approach" (Denis Noble, "Claude Bernard, the First Systems Biologist" (2008), 16. See note 58.) In a 2008 article, he provides "ten principles of systems biology" (Denis Noble, "Mind Over Molecule: Activating Biological Demons," Prologue, in *Control and Regulation of Transport Phenomena in the Cardiac System*, vol. 1123 of *Annals of the New York Academy of Sciences* (2008): xi–xli.). Noble's work pushes present heuristics, and the systems approach generally is part of ongoing advances in biology and medicine. But, as will emerge in reflections below, systems biology struggles within present-day disorientations.

[66] This focus is prevalent in contemporary systems biology. See, for example, references in note 65.

and Walhout,[67] is "Cells are Computers." The second section of Chapter 8, by Lee et al. is "The Cell is More Than the Sum of its Parts," in the first paragraph of which we find the following:

> The new approach has taken shape as 'systems biology,' and is based on the concept that by measuring the behavior of every 'part' of the cell, a coherent set of systems-level properties of cellular behavior will emerge. Systems-level approaches have been defined and shaped by genomics, whereby the cellular state is reported as a single system-wide profile comprised of individual readouts for every gene, protein, metabolite or other molecular element in the cell. ... the data from all available perspectives should be integrated in order to account for the interactions and dependencies between molecular elements vital to cellular functions. By integrating these diverse systems datasets, it is the hope that the ultimate goal of systems biology – the ability to both understand and predict the cellular response to perturbation – will be realized.[68]

The second section of Chapter 16, by Mariottini and Iyengar, is "Cellular Signaling: Pathways to Networks."[69]

That section begins:

> Intrinsic to cellular processes is the receipt and processing of information. The ability of a mammalian cell to live depends on its ability to receive information and respond to a constantly variable environment. Information received through a pathway can often evoke a cellular response. ... Networks result from interconnections between signaling pathways.[70]

However, as already mentioned, while there is an abundance of data on chemical properties, data on alleged cellular signaling networks, as networks *per se*, are not experimental.[71]

---

[67] Martha L Bulyk and A. J Marian Walhout, "Gene Regulatory Networks," ch. 4 in, *Handbook of Systems Biology*, 65.

[68] Anna Y. Lee, Gary D. Bader, Corey Nislow and Guri Giaever, "Chemogenomic Profiling: Understanding the Cellular Response to Drug," ch. 8 in, *Handbook of Systems Biology*, 154.

[69] Chiara Mariottini and Ravi Iyengar, "System Biology of Cell Signaling," ch. 16 in *Handbook of Systems Biology*, 311.

[70] Chiara Mariottini and Ravi Iyengar, 311-312.

[71] See paragraph containing notes 49, 50, 51 and 52. For some details on visualization techniques, see, for example, *"GRN's: Visualization,"* Bulyk and Walhout, 66. Note that 'GRN' is an acronym for "Gene Regulatory Network."

Let us now look to one more example from the literature:

> The subdivision of natural entities into systems is an abstract construct. Systems *per se* do not really exist in reality; rather, they are defined as a set of elements interacting over time and space. Systems theory denotes the transdisciplinary investigation of the abstract organization of phenomena, independent of their substance, type, or spatiotemporal scale of existence. The goal of systems theory is to study emerging properties arising from the interconnectedness and complexity of relationships between parts. Such theory argues that however complex or diverse a system is, there are always different types of organizational structures present, which can be represented as a network of information flow. Because these concepts and principles remain the same across different scientific disciplines such as biology, physics, or engineering, systems theory can provide a basis for their unification. The systems view distinguishes itself from the more traditional analytic approach by emphasizing the concepts of system– environment boundaries, signal input– output relationships, signal and information processing, system states, control, and hierarchies. Albeit systems theory is valid for all system types, it usually focuses on complex, adaptive, self-regulating systems which are termed *"cybernetic."*[72]

On what grounds, however, can systems biology suppose being able to measure "every 'part' of the cell,"[73] and similarly, assert that an "ultimate goal (is) to be able to predict the cellular response to perturbation?"[74] Or, again, on what grounds can it to be asserted that "the pluralism of causes and effects in biological networks is better addressed by observing, through quantitative measures, multiple components simultaneously and by rigorous data integration with mathematical models?"[75] Does it not become increasingly obvious that the variously named 'signaling networks,' 'DNA machinery' and the like, are, in fact figures and figments of the imagination?

Note also that the aims of systems biology are out of sync with contemporary methods in physics, biophysics and biochemistry, all of

---

[72] Melanie Boerries, Roland Eils and Hauke Busch, "Systems Biology," ch. 1 in, Robert A. Meyers, ed., *Systems Biology, Current Topics from the Encyclopedia of Molecular Cell Biology and Molecular Medicine* (Somerset, NJ: John Wiley & Sons, 2012), 7. Italics in source text.

[73] See note 68.

[74] See note 68.

[75] Uwe Sauer et al., "Genetics: Getting Closer to Whole Picture." *Science* 316 (5824): 550-551.

which involve complex layerings of randomness and empirical probabilities. The aims also are out of sync with how biologists work and report on experimental results. In particular, measurements of "every part"[76] of a cell are not available, let alone "simultaneously."[77] And there is nothing available that is anything like a "readout."[78]

The present purpose is not to provide solutions to these difficult problems. Progress will be a community achievement. Nor is it to suggest that systems biology has not been making important advances in biology and medicine.[79] Still, to end this section, it is possible to make an observation about methods: In systems biology, and in contemporary philosophy of biology, a regular emphasis is on: imaginable spatial representations; analogies with computers and machines; and logical structures and conceptual constructs that cannot be verified in living organisms, even when they are one-celled organisms. Evidently, something more is needed. What that more is, of course, is part of the present great challenge to develop empirical method.

2.3 Bird Anatomy

Increasing sophistication within contemporary philosophy of biology, along with the extraordinary reaches of contemporary biology, all bear witness to "the amazing complexity of living organisms,"[80] including the living organisms who are the biologists and philosophers of biology. At this stage, let us enter into elementary empirical searchings, fledgling inquiry that calls for (preliminary) self-description in understanding the adult *columbidae*.

---

[76] See note 68.
[77] See note 75.
[78] See note 68.
[79] See note 65.
[80] *Handbook of Systems Biology*, xi.

2.3.1   *Anatomy of an adult pigeon*

One of many books on pigeon anatomy is *Laboratory Anatomy of the Pigeon*[81] (*LAP*). This book is referenced, not because it is the most up-to-date, or most complete, but because it is accessible to a broad audience, and is easily available. Some such book will be helpful; and if possible, some experience in a laboratory.

In *LAP*, the basic anatomy of the (whole) pigeon is spread out in nine chapters of annotated dissection experiments, with illustrations and photos. The chapters of *LAP* are: 1. External Anatomy and Skin; 2. Skeletal System; 3. Musculature; 4. Digestive System; 5. Respiratory System; 6. Circulatory System; 7. Urogenital System; 8. Nervous System; 9. Special Sense Organs. Of course, a list is no replacement for doing anatomy. But, a small selection of topics from *LAP*[82] can help give some impression of the wondrous complexity of basic pigeon anatomy.

1. External Anatomy: Includes: head, elongated beak, sheath on upper jaw, forelimbs for flight, hind limbs for bipedal terrestrial locomotion and for perching. In more detail: crown, forehead, cere, …, primary remiges, …, tarsus, toes. Dorsal view includes: forearm, wrist, hand, propatagium, …, thigh, tail, ankle, foot {metatarsus, digits}, … . Skin: scales on legs and feet, claws and beak, down feathers, contour feathers, filoplume feathers. Different feather types are found in *tracts* such as capital, spinal, lateral. Various views are given: dorsal view, lateral view, ventral view.

2. Skeletal system: The bones are hollow, contain air spaces, and are often fused together. The arrangement permits strength without weight. There are several pages of details. Figure 18 of *LAP* produces an image for the lateral (dorsal) and medial (ventral) views of the wink skeleton. These include fused carpals, metacarpals, and much more.

3. Musculature: A highly complex weave of 175 muscles.

4. Digestive system: Starting with mandibular beak, tongue and oral cavity, this part of the dissection includes: body cavity and all viscera such as heart, liver lobes, intestines, membranes that support viscera, ligaments, and various compartments.

---

[81] Robert B. Chiasson, *Laboratory Anatomy of the Pigeon*, 3rd Ed., Dubuque (Iowa): W. C. Brown, 1984. Hereafter called *LAP*.
[82] See note 81.

5. Respiratory system: This is a complicated system of various air sacs in the body cavities and air spaces inside many bones. Other parts include: larynx, trachea, syrinx, tempaniform membranes, bronchi, lungs.

6. Circulatory system:

This includes the heart, with its atria, ventricles, veins, arteries, aortic arch, chambers, cup like flaps fused at their borders with adjacent flaps. Also included are the vast vein works with flow patterns of the whole body that in particular, include an intricate mesh of wing veins called radial, dorsal, ulnar, brachial, auxiliary veins, and several others. Major veins can be found and distinguished throughout the body, connecting to all organs, limbs, tracts.

7. Urogenital System.

8. Nervous System.

9. Special Sense Organs.

2.3.2 *Data of pigeon anatomy*

There is data of seeing, touching, hearing and smelling pigeons. And, from ancient times to present, for some there are even data of how pigeons taste. In biological science, though, it is through anatomical studies and dissection that one obtains initial presentations within sense experience. And, normally, investigators go on to make re-presentations of initial presentations. So it is that, in anatomy texts, one finds photos and other products of imaging technologies; extensive and intricate diagrams; variously organized re-presentations for parts, tissues, membranes, body fluids; and descriptions of connectivities between all of these. Even an introductory pigeon anatomy text such as *LAP* requires more than a hundred pages of increasingly complex images, diagrams, charts, representations, names and symbols. And, within a basic position, one grows in being attuned to the fact that all of these images are within one's own sensitive consciousness.

2.3.3 *Some insights in pigeon anatomy*

When studying anatomy, one does not simply inspect once, but does so from many different perspectives. There is also an effort to organize. Basic pigeon anatomy, therefore, includes vast and indefinitely large aggregates of descriptive insights; more, or less, organized aggregates of descriptions - of a pigeon and its many inspected and dissected parts.

Through preliminary self-attention, one may self-notice that, within one's immediate field of inspection, inspected parts, membranes, fluids and the like (as the case may be) are descriptively connected within one's field of inspection. But, as one moves from one instance to another, loci of attention within sensibility also are connected in another way, that is, within one's thinking about the whole pigeon. "What is *it*"? The 'it' includes all described and descriptively connected parts - present, re-presented and remembered. At any one time, what is present within one's imagination or view is only one aspect of, or one perspective on, a pigeon's anatomy. But, as long as one is doing pigeon anatomy, then, as in the *LAP*, descriptions are understood to be about a whole pigeon.

2.3.4 *Cell line, adherens, the extra-cellular matrix and connective tissues*

In dissection, one uses various instruments. Examples include the hemostat; forceps with serrated teeth; scissors with one blunt tip; scalpel with replaceable blades; probe with blunt tip; elementary magnification devices; as well as more sophisticated technologies such as light microscopes (compound, phase contrast, dark-field), fluorescent microscopes and electron microscopes.

Contemporary technologies reveal fine descriptive differences in tissues and body parts. For example, one finds tissue patternings called *cells*, as well as somewhat similar patternings interior to cells. Many of the patternings interior to a cell are called *organelles* (such as *mitochondria*). There are also *adherens* (what eventually are understood to be inter-cellular proteins); many types of fluids; varieties of extra-cellular matrix in various anatomical locales; and so on. Note that this includes described components of the avian brain and neural mesh - "central" (visibly within skull and spinal cord) and "peripheral" (outside of the brain and spinal cord) nervous "systems" (CNS and PNS respectively); and an "Autonomic Nervous System" (ANS) by which (it is found) basic visceral functions and thermoregulation are auto-regulated. In other words, techniques and technologies of contemporary science allow us to obtain subtle and increasingly nuanced descriptive differentiations of the whole pigeon and its many describable parts.

Altogether, then, in contemporary anatomy of the pigeon, one thinks about a whole pigeon, and inspects many described parts. One has presentations, representations, diagrams, names and symbols, combined and juxtaposed in diverse and complex ways, according to focus. Some of these are called "external features" of the organism, such as featherings, limbs, and so on[83]; and others are called internal features, such as organs, anatomical parts; cell lines and organelles; various types of extra-cellular matrix and *adherens*.

Again, note - or rather, self-note - that all of these described features are connected and differentiated within one's imagination; and all of what one describes and imagines, one holds together within a unifying question-poise: "What is a pigeon?" However, as discussed in the next section, descriptive understanding attained in anatomical studies is, of course, only part of contemporary understanding of *columbidae*.

2.4 The Chemically Talented *Columbidae*

2.4.1 *Flight chemistry*

Today, the science of biochemistry has omni-disciplinary reach, and new results are published weekly. An early major breakthrough in biochemistry was the discovery of the tricarboxylic acid cycle (TCA), or Krebs cycle. The TCA cycle was discovered in the 1930's by Hans Adolf Krebs (1900-1981), with the help of his assistant Leonard Victor Eggleston (1920-1974). [84] They were investigating the metabolic activity of pigeon wing muscle, muscle crucial for avian flight. Prior to the discovery, the

---

[83] See Section 2.3.

[84] See Figure 2.1. Krebs was awarded a 1953 Nobel Prize for his discovery of the TCA cycle, also now known as the Kreb's Cycle. See Hans Adolf Krebs and Leonard Victor Eggleston, "The oxidation of pyruvate in pigeon breast muscle," *Biochem. J.* 34 (3), (March, 1940): 442- 459. In his own words, Krebs' journey was a "slow" creative journey, "extended over some five years beginning ... in 1932." See the first paragraph of: H.A. Krebs, "The History of the Tricarboxylic Acid Cycle," in *Perspectives in Biology and Medicine*, Vol. 14 (1970): 154-170.

community was looking into how certain organic (di- and tri- carboxylic) acids are oxidized by animal tissues. The Krebs cycle partly explains consumption of O_2, as well as the metabolic product CO_2.

This sub-section is to help us get some slight impression of the magnificently complex biochemistry of *columbidae*; and, that way, also obtain data in that understanding of the biochemistry of *columbidae*.

Diagrams for the TCA cycle are easily available in textbooks, or online. Some are more complex than others, depending on context. For example, see Figure 2.1. The TCA cycle is a series of reaction equations.[85] The series, however, circles back to the first reaction equation: Some products of the first equation are reactants for the second; some products of the second are reactants for the third; and so on. Finally, some products of the penultimate equation are reactants for the initial equation – hence, the name, "cycle."

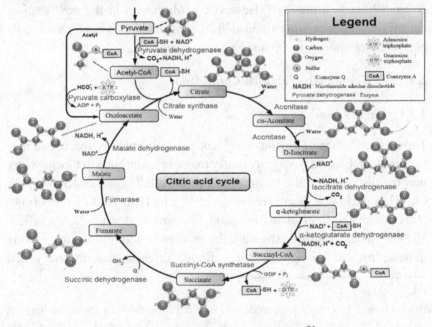

Fig. 2.1. Tricarboxylic acid cycle (TCA).[86]

---

[85] See note 88 for a description of the entire TCA cycle.

[86] Wikipedia contributors, "Citric acid cycle," *Wikipedia, The Free Encyclopedia*, https://en.wikipedia.org/w/index.php?title=Citric_acid_cycle&oldid=670345723.

Verification of the TCA cycle can be obtained through well-known experiments, going back to the original work done by Krebs and his collaborators. Minced fresh pigeon wing muscle is suspended in a phosphate saline solution. (The solution is found to not inhibit metabolic activity of fresh muscle tissue.) Then, through combinations of sequences of *in vitro* (and for the pigeon, *ex vivo*) experiments, each of the reactions of the full TCA cycle can be obtained.

Here, let's look to just one of those reaction equations. For muscle suspended in solution, one of the reaction equations is:

$$\text{fumarate} + \text{pyruvate} + 2O_2 \rightarrow \text{succinate} + 3CO_2 + H_2O.$$

Chemical formulas for the acids are: fumarate, $HO_2CCH{=}CHCO_2H$; pyruvate, $CH_3COCOOH$; and succinate, $HOOC\text{-}(CH_2)_2\text{-}COOH$. Experiments reveal that fumarate acts as a catalyst: O_2 uptake increases with an increased supply rate of fumarate, and decreases when the supply rate of fumarate is reduced. Similarly, the production rates of succinate, CO_2 and H_2O depend on the presence or absence of fumarate and pyruvate.

For the reaction equation just given, what are one's data, and one's understandings? Of course, to follow up with these questions, one needs to have some familiarity with experimental arrangements. One also needs to have some competence within the relevant bio-chemistry, the full context of which is a chemical system in which elements and compounds are mutually defined through a vast matrix mesh.[87]

What, then, do the TCA cycle equations tell us about the minced fresh muscle tissue? It may help to draw attention to two aspects of the minced muscle tissue experiments. The chemical reaction equation is verified when (i) the muscle tissue (ii) is in a non-inhibitory solution. Being a little more precise, we could write: *When suspended in a non-inhibitory phosphate saline solution,*

$$\text{muscle tissue} + \text{fumarate} + \text{pyruvate} + 2O_2 \rightarrow \text{muscle tissue} + \text{succinate} + 3CO_2 + H_2O.$$

Now, does this not mean that muscle tissue has chemical properties? For, *in vitro*, fresh muscle tissue has the capacity to oxidize in the way expressed by the reaction equation just given.

---

[87] This is sliding past challenging foundational exercises in chemistry that eventually will be normalized within education in science and philosophy of science. See Section 4.4.

But, the full TCA cycle is not just this one equation. And, the other equations are verified in similar ways. We find, then, that *in vitro* and *ex vivo*, fresh minced wing muscle tissue has the capacity to chemically react in several ways, through a cycle of intertwined chemical reaction equations. Each equation of the cycle is verified through its own combinations of apparatus and reactions. Various tissue samples are prepared, suspended in different solutions at various controlled temperatures. Reactants are introduced into solutions, yielding different products. What, though, might these fragmentary results have to do with a living adult pigeon, *in vivo* and *in situ*?

2.4.2   Chemical talents of a living pigeon

The aim of this chapter includes making progress toward a preliminary heuristics (or, rather, as will be seen by the end of the chapter, what would be better called a preliminary *pre*-heuristics) for holding these results together, not as a pattern of distinct chemical reactions *in vitro* and *ex vivo*, but as they pertain to a living pigeon, *in vivo* and *in situ*. One obvious difference between *in vitro* and *in vivo* results can be seen easily by noticing that for *in vitro*, the different TCA intermediates are introduced into various combinations of *in vitro* solutions by an investigator. The living bird, though, survives without any such interference.

What, then, do results about TCA intermediates have to do with a *living* bird? Contemporary biology asserts both that: (i) the TCA intermediates (complex biomolecules) are "inside" the living pigeon[88]; and (ii) that while

---

[88] In contemporary descriptions of the TCA cycle, note references to molecules and reactions (allegedly) inside the organism. For example: "Tricarboxylic acid cycle: the second stage of cellular respiration, the three-stage process by which living cells break down organic fuel molecules in the presence of oxygen to harvest the energy they need to grow and divide. This metabolic process occurs in most plants, animals, fungi, and many bacteria. In all organisms except bacteria the TCA cycle is carried out in the matrix of intracellular structures called mitochondria. The TCA cycle plays a central role in the breakdown, or catabolism, of organic fuel molecules—i.e., glucose and some other sugars, fatty acids, and some amino acids. Before these rather large molecules can enter the TCA cycle they must be degraded into a two-carbon compound called acetyl coenzyme A (acetyl CoA). Once fed into the TCA cycle, acetyl CoA is converted into carbon dioxide and

net concentrations tend to stabilize, the various TCA intermediates are in flux patterns with relative rates approximated by *in vitro* reaction equations of the TCA cycle. In contemporary scientific literature one even finds sophisticated computer generated 3-D imaging of proteins and other complex biomolecules, represented in various folded or unfolded states called *conformations*.[89] Examples of the geometrical representations are easily available in the literature and online. Figure 2.2 is a diagram for what are called primary, secondary, tertiary and quaternary conformations:[90]

Biomolecules are not only said to be imaginably inside organisms, but as imagined in such geometric diagrams. (Older style examples included "atom and rod" diagrams, such as in Figure 2.1.) Recent advances in experimental and computational methods provide even more nuanced figures of macro-molecules, their various chemical and transitional states between primary, secondary, tertiary and quaternary conformations.[91]

So, before thinking further about the living bird, let us pause over the common claim that biomolecules are "inside." To do that, let us look again to *in vitro* results. As already discussed, verification *in vitro* of the TCA cycle involves multiple sets of tissue samples, in various sequences and

---

energy. ... The TCA cycle consists of eight steps catalyzed by eight different enzymes. The cycle is initiated (1) when acetyl CoA reacts with the compound oxaloacetate to form citrate and to release coenzyme A (CoA-SH). Then, in a succession of reactions, (2) citrate is rearranged to form isocitrate; (3) isocitrate loses a molecule of carbon dioxide and then undergoes oxidation to form alpha-ketoglutarate; (4) alpha-ketoglutarate loses a molecule of carbon dioxide and is oxidized to form succinyl CoA; (5) succinyl CoA is enzymatically converted to succinate; (6) succinate is oxidized to fumarate; (7) fumarate is hydrated to produce malate; and, to end the cycle, (8) malate is oxidized to oxaloacetate. Each complete turn of the cycle results in the regeneration of oxaloacetate and the formation of two molecules of carbon dioxide" ("Tricarboxylic acid cycle," *Encyclopædia Britannica Research Starters*, 2013. Accessed May 4, 2014.).

[89] See, for example, Michael J. Sternberg, *Protein Structure Prediction, A Practical Approach*. Oxford: Oxford University Press, 1996/2002. José Nelson Onuchic and Peter G Wolynes, "Theory of protein folding," *Current Opinion in Structural Biology* vol. 14, issue 1 (February 2004): 70-75. Alka Dwevedi, *Protein Folding: Examining the Challenges from Synthesis to Folded Form*. Cham: Springer International, 2015.

[90] See note 92.

[91] See, for example, Rossitza N. Irobalieva, Jonathan M. Fogg, Daniel J. Catanese, et al., "Structural diversity of supercoiled DNA," *Nature Communications* vol. 6, Article number: 8440 (October, 2015).

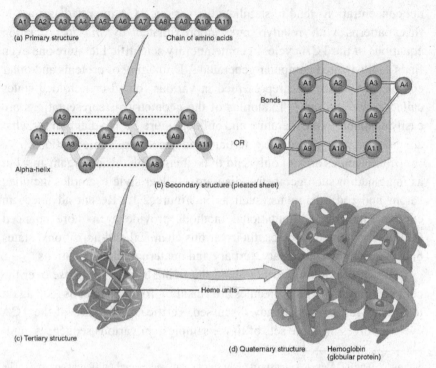

Fig. 2.2. Primary, Secondary, Tertiary and Quaternary Structures of Amino Acid Residues.[92]

---

[92] The example shown here is hemoglobin, a protein in red blood cells which transports oxygen to body issues (https://en.wikipedia.org/wiki/Protein#/media/File:225_Peptide_Bond-01.jpg). "The primary structure is the one-dimensional sequence of amino acids in the protein or peptide. ...The secondary structure is a regular structural unit formed by sequence regions in a protein (from a sequence perspective: two-dimensional). Basically, these are of two types: α-helices and β-sheets. There are also a number of less common structural motifs that sometimes are regarded as secondary structures (e.g., β turns, left-handed helices). ... Tertiary structure is the three-dimensional fold of the protein, or how the secondary structure elements are packed against each other. ... Quaternary structure is the way several polypeptide chains together can form a functional complex. In many cases more than one chain is needed to achieve a particular function, e.g., hemoglobin with four subunits or the ribosome that is put together by a large number of separate protein and RNA molecules" (Henrik Hansson; Gunnar Berglund & Evalena Andersson, *Introduction to Swiss Pdb Viewer: The structure levels in proteins*. Uppsala Universitet (2003); edited 2006: Mats Sandgren, Uppsala Universitet; completed 2007-2011: Maria Selmer, Uppsala Universitet. http://xray.bmc.uu.se/kurs/BSBX2/practicals/practical_2/practical_2.html).

combinations of sequences of *in vitro* chemical break-down reactions. Products obtained are combined with further reactants, as well as products of other reactions, all under the control of understanding in biochemistry. Samplings of products obtained are isolated, and further analyzed by centrifuge, mass spectrometry and/or charge spectrometry for isotopes. Altogether, the various experimental procedures help us verify physical and chemical properties of reactants and products from various combinations and sequences of reactions, derived from multiple muscle tissue samples in solution.

With regard to (i) above, it is to be noted and self-noted that in *in vitro* experiments, one is not observing individual compounds or conformations. It is true that there are ongoing advances in imaging technologies that now, for example, can provide images of individual biomolecules as topographic filaments and shapes.[93] But, observe and self-observe that what we verify in chemical analysis of tissue samples are not images or diagrams of individual molecules, but chemical reaction patterns of (multiple) tissue samples, along with structured sets of secondary, tertiary, quaternary, ..., reaction patterns of other reactants and products.

Yet, (i) and (ii) together say more. For the assertions of (i) and (ii) are not just that properties of various TCA intermediates can be verified for *in vitro* tissue samples, but that they are properties of a living pigeon, *in vivo*. To be sure, the *in vitro* results are compelling. But, what evidence is there for the assertion that the TCA cycle functions in a living bird?

What follows next is a pointing to some of what is known about the *living* pigeon: (1) Complementing the fact that *in vitro* reaction rates are all mutually comparable, it is found that *in vivo* production rates of CO_2 and consumption rates of O_2 closely compare with *in vitro* rates. (2) If

---

[93] See, for example: Q. Huan et al., "Spatial imaging of individual vibronic states in the interior of single molecules," *Journal of Chemical Physics*, vol. 135, issue 1 (July 2011): p. 014705. 6p; and H. Inagawa at al, "Reflecting microscope system with a 0.99 numerical aperture designed for three-dimensional fluorescence imaging of individual molecules at cryogenic temperatures," *Scientific Reports*, vol. 5 (2015): 12833. What is the significance of experimentally obtained images of individual biomolecules? Basic here still is the need for control of meaning. Advancing within a basic position, one distinguishes components of realities intended. See Section 4.4.

either or both oxygen and carbohydrate supplies to the living pigeon are reduced, muscle function begins to fail. (3) If selected TCA intermediates are injected directly into local tissue or bloodstream, muscle activity can be catalyzed, in patterns and rates compatible with TCA *in vitro* results. (4) In some *in vivo* experiments, glucose and oxygen supplies to selected muscles are accurately controlled; blood plasma is then extracted from a living pigeon at time intervals of 1 second, 2 seconds, ..., 6 seconds, and so on — that is, at time intervals comparable to TCA cycle *in vitro* rates. Concentrations of intermediates in the subsequent blood samples can be compared with *in vitro* results. It is found that *in vivo* muscle activity, flux rates in glucose and oxygen levels, concentrations and flux patterns of TCA intermediates present in extracted plasma are all comparable with estimates based on the *in vitro* experiments.

Other results from biochemistry reveal further biochemical sophistications. For example, in a living pigeon, the TCA cycle is understood to be "chemically downstream" from another (non-cyclic) chemical sequence called *glycolysis*. In this context, "chemically downstream" means that products of *glycolysis* are understood to be reactants for the TCA cycle. See Figure 2.3. More generally, the expression "chemical pathways" is used when chemical sequences, cycles, or other biochemical connectivities are known. And, at present, there are more than 500 known chemical reaction pathways in a typical pigeon cell. Although, "almost all pathways lead to the TCA cycle."[94]

For yet another example of biochemical pathways and connectivities, one may recall that oxygen is said to be carried to cell sites for the TCA metabolic cycle. A binding of oxygen is said to take place in the lungs, across what are described as membranes of small balloon-like structures called *alveoli* attached to branches of the bronchial passages. Oxygen is said to bind to protein "inside" red blood cells. A special form of hemoglobin produced in this way is *oxyhemoglobin*, formed during respiration when oxygen binds to the *heme* component of the hemoglobin protein component in red blood cells. Oxygen is then transported to cell sites so that it may then function in the oxygen dependent TCA cycle.

---

[94] David R. Dalton, *Foundations of Organic Chemistry, Unity and Diversity of Structures, Pathways and Reactions*. Hoboken, NJ: John Wiley and Sons Inc., 2011.

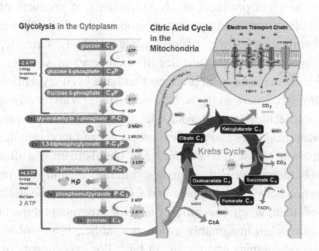

Fig. 2.3. Major pathways of cellular respiration in typical eukaryotic cell.[95]

In another direction, carbon dioxide produced by the TCA cycle binds to de-oxygenated hemoglobin (which is then called *carbinohemoglobin*) so that it can be transported back to the lungs for removal from the organism by being released into the environment.

As progress in biochemistry continues to show, diagrams, charts and maps of multi-node pathways can be extremely useful as images.[96] However, such images are not re-presentations of visible or experimentally observed pathways. Just as we enquired into whether or not there is data on TCA intermediates inside cells *per se*, here too, we can ask: What evidence is there, if any, for asserting that hemoglobin protein molecules, in any conformation, are imaginably inside red blood cells?

It can again help to look again to *in vitro* results. In that case, proteins are said to be "inside" red blood cells when blood samples chemically function appropriately: that is, when blood samples can be chemically

---

[95] Wikipedia contributors, "Cellular respiration," *Wikipedia, The Free Encyclopedia*, https://en.wikipedia.org/w/index.php?title=Cellular_respiration&oldid=672310324 (accessed July 29, 2015). The *Wikipedia* reference is given as a convenience for the reader. The pathways are well known and can be found in most undergraduate textbooks in biochemistry.

[96] St. Thomas Aquinas, *Summa Theologiae* I, q. 84, a. 7.

processed to yield products or crystallizations of products of chemical reactions that are secondary, tertiary, and so on, that way verifying the many chemical properties of said proteins. And, normally, this also includes spectral analysis (physics of chemistry) of samples, as a help toward identifying chemical products that have been crystallized. In other words, through *in vitro* analysis, here too there is no *inspection* of networks of macro-molecules. Instead, what one discovers and verifies are networks of chemical properties and physical properties of blood samples.

What about the shift to *in vivo*? The bloodstream is a fluidic part of what can be described as the pigeon's "internal anatomy."[97] Among other things, ratios between O_2 uptake and CO_2 release can be determined. In both cases, then, whether *in vitro* or *in vivo*, experimental methods do not reveal hemoglobin imaginably inside, but, rather, *hemoglobin properties* of both blood samples and the living bird. For experiments *in vitro*, one discovers and verifies chemical properties of *in vitro* samples; while for experiments *in vivo*, inquiries necessarily are about living pigeons, and one discovers and verifies chemical properties of a living pigeon, in a lab, or sometimes even *in situ*.

Two main types of experiments are *in vitro* and *in vivo*. But, they are not unrelated. For both are part of the effort to understand the living bird. What, then, do the various questions and results pointed to in this section tell us about a living pigeon? Evidently, the living pigeon has a vast and complex range of chemical properties. More data could be obtained if, for example, we were to go on to explore some of the extraordinary subtleties of oxygen binding and carbon dioxide release, or enter into details of some of the several hundred chemical pathways that have been verified in the pigeon. For now, enough preliminary work has been done to intimate the future possibility of stating with a control of meaning that, truly, a living pigeon is a bird of many chemical *talents*![98]

---

[97] See Section 2.3.

[98] *talent*: archaic: a characteristic feature, aptitude, or disposition of a person or animal. Earlier, from Proto-Indo-European, tele- "to lift, support, weigh." For a glimpse into the extraordinarily complex repertoire of avian chemical talents, see Lewis Stevens. *Avian Biochemistry and Molecular Biology*. The book contains more than 200 pages of dense technical summary of biochemistry particular to birds. Most paragraphs point to, and draw on vast tracts of results in the scientific literature. And, "(a)reas where there are only slight

2.5 Physics; and Chemistry of an Adult Pigeon

2.5.1 *The physically; and chemically talented pigeon*[99]

In the previous section, the emphasis was on *chemical* properties. But, as already alluded to, there are also other properties of organisms, known, for example, through physics. In particular, there is spectral analysis of tissue samples. The focus of this chapter is not bio-physics, as such. Still, a few comments here will help toward filling down the emerging pre-heuristics for a living pigeon. In fact, this also will help prepare the ground for thinking about *higher* properties. Note, however, that heuristics for what is to be meant by *higher* are yet to be determined.

One may recall that fluids obtained from a pigeon's anatomy have many of the physical properties of Brownian aggregates[100] and statistical mixtures. More, though, is known. For instance, *in vivo*, there is uptake of carbon isotopes when they are injected into tissues sites of a living pigeon. Nuclear magnetic resonance apparatus aimed at anatomical sites can track metabolic functioning of labeled acetates. Fluctuation rates (as well as stabilization patterns) for the spectra (physics) of TCA intermediates are found to be in accord with estimates based on *in vitro* TCA cycle chemical reactions obtained through appropriate sequences of prepared reaction baths.

But, physics also contributes to understanding the physiology of the whole organism *in situ*. For example, an adult Rock Pigeon has an approximately constant mass (typically somewhere in the range 0.2kg - 0.4 kg). And, biomechanics and aerodynamics partly explain functionalities of various skeletal formations, musculature patterns; feather formations; wing shape; and lift.[101]

---

differences between birds and mammals, and which are well covered in standard textbooks of biochemistry and molecular biology, either are not considered or discussed only briefly" (Lewis Stevens, xi).

[99] The significance of the semi-colon begins to emerge in the work below. For now, it can be taken as referring to the fact that biology finds pigeons have properties of both kinds. In an expanded context, this is revisited in Chapter 4, in discussion of the one-celled organism *E. coli*.

[100] A study of Brownian motion goes well beyond introductory discussion about statistics and probability given in Chapter 1.

[101] A point of entry into the literature is, Anders Hedenström, "Mechanics of Bird Flight:

Evidently, an empirically grounded pre-heuristics for a living pigeon will need to include some kind of layering of physics with chemistry. In other words, based on present knowledge, a living pigeon is both physically; and chemically multi-talented.

2.5.2 *"When (the pigeon flies) the whole (pigeon flies)"*

A living pigeon (with its TCA cycles, chemical pathways and physical properties) survives not through the assistance or interference of investigators, but while it goes through its various activities *in situ.* It flies; breathes oxygen; moves about in its environment to obtain food stuff; sleeps; preens its own feathers and spreads oils from a tail gland. Pigeons mate, and nest in locations too high for most ground predators to reach. Generally, the pigeon is a social animal among its own kind. At times it is mildly aggressive with other pigeons. And, at other times a pigeon takes flight, if, for example, a predator's moving shadow is detected. And so on.

With these things in mind, one may push a little further toward a *pre-heuristics*, by including these further described capacities. For now, though, let's start with just one of the activities mentioned, the one that so regularly catches attention, namely, *flight.* Certainly, something quite marvelous happens. Re-coining an expression about babies laughing, "When a pigeon flies, the whole pigeon flies!" [102]

The fact is that, during *flight* (or any other activity), all of the anatomically, physically and chemically differentiated parts of a pigeon function together. How, though, can this be written down? Keeping the real pigeon in mind, might we not start by labeling the heart H, the lungs L, bloodstream B, wing muscles W, stomach S, and so on, leading to a symbolism that matches known anatomical differentiations of a pigeon?

But, as is known within contemporary biology, each organ has several cell lines, each with known differences in chemical functionings. If we

---

The Power Curve of a Pigeon by C. J. Pennycuick," *The Journal of Experimental Biology* (May 15, 2009): 1421-1422.

[102] Inspiration for the title of this section was from a response given by Bernard Lonergan, during a question and answer session: "When the baby laughs the whole baby laughs" (Bernard Lonergan, *Philosophical and Theological Papers* 1965-1980, vol. 17 in, *Collected Works of Bernard Lonergan* (Toronto: University of Toronto Press, 2004), 150).

keep to present symbolism, we would need to include something like H1, H2, ...; L1, L2, ...; B1, B2, ...; and so on, for the various cell lines of the pigeon heart, lungs; bloodstream; ..., respectively. Note, too, that the anatomical, physical and chemical connectivities (chemical pathways, cycles and other schemes of recurrence within a pigeon) are diverse, complex, and multiply link large aggregates of parts within the organism. The temporary symbolism H1, H2, ...; L1, L2, ...; B1, B2, ..., quickly becomes unwieldy. Moreover, results still are fragmentary; and what is lacking is a symbolism for a whole bird in *flight*.

Notation can be improved somewhat by using a generic 'P' to represent a (described) anatomical part of the pigeon – descriptively, physically and chemically differentiated within the anatomy. There are, of course, many anatomical parts. For a pre-heuristics, therefore, we will need to introduce at least some kind of indexing. For now, we can use 'a,' pointing loosely to the fact that there are various parts of the anatomy.

Each anatomical part P(a) has physical and chemical properties, cycles and pathways $(p_i;c_j)_a$. For example, in cell populations, the notation $(p_i;c_j)_a$ includes the intra-cellular TCA cycle, the *glycolysis* pathways, as well as physical spectral flux patterns; for eukaryotic cells, there is regulation of charge gradients across membranes of mitochondria. The biochemistry of pluripotent cells is not the same as that of differentiated cells. And so on. Combing all of this physics and chemistry into our notation for anatomical parts P(a), we get functionalities $(p_i;c_j)_a$.[103]

Now, for the living pigeon, *in situ*, lungs allow for a chemical binding of oxygen to the bloodstream of the organism; the bloodstream is needed to transport bound oxygen to cell sites, where aerobic TCA cycles are verified in muscles, organs and other cell lines. Recall also from *LAP*[104] that there are various complex so-called systems in the anatomy: External Anatomy and Skin; 2. Skeletal System; 3. Musculature; 4. Digestive System; 5. Respiratory System; 6. Circulatory System; 7. Urogenital System; 8. Nervous System; and 9. Special Sense Organs. None of these systems are independent systems. None function separately from the

---

[103] Regarding symbolism, for the moment I am simply appealing to known anatomy, biophysics and biochemistry. Regarding symbolism, see also note 114, below.
[104] See note 81.

whole organism. All work together, through vast arrays of connective networks of physical-chemical gradients, flows, pathways and cycles.

One way to point to all of this, heuristically, is to use $(p_i;c_j)_{(a,b)}$ to represent physical and chemical connectivities that functionally connect $(p_i;c_j)_a$ and $(p_i;c_j)_b$, for anatomical parts P(a) and P(b) respectively. Again, though, during flight, the whole bird flies. All anatomical parts P(a) function together. For the time being, then, we can use spanning brackets $< *, * >$ to represent the living dynamic unity-identity-whole[105] *in flight*.

Brining all of this together, a pre-heuristics for flight is obtained: That is, an *whole-organism* physical and chemical metabolic profile for flight:

$$(flight) = <P(a): (p_i;c_j)_a, (p_i;c_j)_{(a,c)}, (p_i;c_j)_a >(flight).^{106}$$

2.5.3 *Flight and other activities of the adult pigeon*

A pigeon does many more things besides fly. And, most of the comments above apply equally well to all of its other activities. For example, there are metabolic profiles for

$$(resting) = <P(a): (p_i;c_j)_a, (p_i;c_j)_{(a,c)}, (p_i;c_j)_c >(resting),$$

$$(mating) = <P(a): (p_i;c_j)_a, (p_i;c_j)_{(a,c)}, (p_i;c_j)_c > (mating),$$

$$(escaping) = <P(a): (p_i;c_j)_a, (p_i;c_j)_{(a,c)}, (p_i;c_j)_c > (escaping),$$

and so on.

Here, our pre-heuristics for the various activities is coherent with verifiable results of contemporary biology. Without needing to impose non-verifiable constructs, the heuristics is explicit in that whatever the

---

[105] See note 55.

[106] Results about migratory bird species are obtained by Mèta M. Landys et al., "Metabolic profile of long-distance migratory flight and stopover in a shorebird," *Proceedings of the Royal Society B* (2005) 272: 295–30. Results for the pigeon are obtained, for example, in David Costantini, Gaia Dell'Ariccia and Hans-Peter Lipp, "Long flights and age affect oxidative status of homing pigeons (*Columba livia*)," *Journal of Experimental Biology*, 211 (Feb., 2008): 377-381. (The species name *columba livia* is derived from the Latin *livor*, for the bluish color of the bird.)

activity is, it is always the whole organism. And so, for each activity, $(activity) = <P(a): (p_i;c_j)_a, (p_i;c_j)_{(a,c)}, (p_i;c_j)_c > (activity)$.

In addition to field studies such as those mentioned in note 106, wind tunnel experiments and analysis of plasma also have confirmed that there are specific levels of fatty acids, glycerol and other acids during extended flight, different from, say, when a pigeon is resting. And, it is known that it is an organism-wide metabolic profile that sustains extended flight. If, for instance, between anatomical parts '3' and '7', say, physical and chemical connectivities $(p_i;c_j)_{(3,7)}$ are found to be within flight parameters, this does not necessarily correspond to *flight*. All parameters must be functioning together, all at once, within their respective metabolic flight ranges.

There is, of course, tremendous flexibility and variation. And, metabolic profiles are not mutually exclusive. A bird may be resting, and eating; flying, and in courtship; and so on. But, as experimental results confirm, there are limits. Not all metabolic profiles correspond to flight; not all profiles correspond to sleep. Some correspond to rest; some to sleep; some correspond to mating; some to predator-escape; and so on.

The pre-heuristics here, for flight and other activities of the pigeon, explicitly point to verifiable anatomically correct organism-wide metabolic profiles. But, as a bird moves through its various activities, metabolic profiles are of average concentrations, and statistical distributions, all of which *in situ* are in rapid flux — locally in anatomical parts, and organism-wide within the whole pigeon. The components of metabolic profiles are understood through biophysics and biochemistry. And, as mentioned earlier, these results are fragmentary, and statistical, even when whole-organism profiles can be reliably matched with activities.

But, there are verifiable regularities in a pigeon's many *in situ* activities. While discoveries of complex physical and chemical pathways within the organism reveal that the pigeon has *physical; and chemical* properties, there are further regularities that (a) presuppose known metabolic profiles; (b) by the same token, are not explained by physical or chemical law; and (c) therefore, invite further inquiry. Evidently, the pigeon is not merely physical and not merely chemical!

This is touching on the need of a heuristics that will include a verifiable multi-layeredness,[107] a not-merely-physics; and not-merely-chemistry. And, that heuristics will need to include what it will mean to say "not merely." [108] For now, though, it will be helpful to gather a little more data on activities of an adult pigeon.

2.6 Pigeon Activities: Below Conscious; and Conscious

In contemporary avian science, it is well known that for the pigeon, there are two main types of biological function: *below-conscious* and *conscious*. Examples of below-conscious functions are: salivation, digestion, perspiration, pupil dilation, blood circulation; hormone production and circulation; urinary functions; integumentary functions of skin cell lines and feathers that protect the body from the environment; and so on.

Below-conscious functions are not strictly below conscious, but generally remain so unless problems arise, such as in illness. It is known that below-conscious functions are regulated through an intricate Autonomic Nervous System (ANS). The ANS divides in various ways, in its anatomical configurations and corresponding functions. There are, for instance, the sympathetic parts of the ANS (functional in fight, inhibits digestion, and so on) and the parasympathetic parts (functional in visceral activities such as digestion). For both sympathetic and parasympathetic functions, there are afferent sensory neurons, and efferent motor neurons. And, of course, there is much more besides.

The below-conscious functions are sustained only so long as appropriate food stuffs are secured, and the pigeon remains in a suitable environment. But, how does food stuff get to the stomach for digestion? Evidently, the pigeon has further capacities. In other words, the pigeon has capacities where it is conscious. And, this points to a further division in types of function.

Some of the below-conscious functions (such as blood circulation, hormone circulation, and TCA cycles) are *intra-organism* (in the case of the TCA cycle, *intra-cellular*). Other intra-organism below-conscious

---

[107] This is a major problem in contemporary science and philosophy of science. See, for example, notes 110 - 111 and 113 -117, below.

[108] These issues will be further explored in Section 2.6.

functions include cell repair and various *corrective schemes* that, for example, remove mutated cells, and also remove compounds that have been metabolized but are inimical to the organism; there are also *defensive schemes*, where the organism chemically neutralizes, segregates or removes toxins and infectious entities that have not yet been metabolized but have penetrated membranes. And, the continued functioning of, for example, the below-conscious TCA cycle, depends not only on an adequate supply of carbohydrates, but also on the pigeon's capacity to capture and bind oxygen from the environment (likewise to remove CO_2 from the organism and release it into the environment).

What is found, then, is that below-conscious functions of the pigeon divide into (i) intra-organism functions; and (ii) intra-environment (capacities relative to its environment). Of course, this division is not a separation in the pigeon, but a functional distinction.

The *conscious functions* primarily are intra-environment, although it is known that respiration, for example, can be both; and that what normally are below-consciousness can become conscious when there is illness or other problems. There a pigeon's five senses: sight, touch, taste, smell and hearing. But, it is thought, now, that pigeons also have a "sixth sense," through its *cere*,[109] by which it is sensitive to the earth's magnetic field orientation (allowing for directional stability in distance flights).

All of a pigeon's conscious capacities reveal flexible and dynamic coordination of senses, special sense organs and specialized parts of brain-centered afferent and efferent central and peripheral nervous systems. A study of conscious functions of a pigeon, and in what sense *conscious functions* are not merely *below-conscious*, would need to draw on results of pigeon neuroscience, psychology and behavioral studies. This, however, would go well beyond the scope of this introduction and invitation to the new empirical method.[110]

---

[109] Cordula V. Mora, Michael Davison, J. Martin Wild and Michael M. Walker, "Magnetoreception and its trigeminal mediation in the homing pigeon," *Nature*, Vol. 432, Nov. 25 (2004).

[110] Replacing "below conscious" with botanical, and "conscious" with zoological, the more precisely named combination becomes (physical; chemical; botanical; zoological), which can be abbreviated to (p_i; c_j; b_k; z_l). For future reference, recall Lonergan's pointers about animal species: Within a developing heuristics of aggreformic entities, "(a)n explanatory

It is possible, though, to make a beginning here toward a further assembling of our pre-heuristics. And, that way, some progress is possible also toward the meaning of the *semi-colon* ';' appearing in sections 5 and 6.

2.7 Supracoracoideus

The pigeon's largest muscles are the breast muscles which provide the powerful wing stroke essential for lift and flight. A different muscle, also crucial for flight, is called *supracoracoideus*. There is one for each wing, a strong muscle that raises the wing between beats. Note that the etymology of *supracoracoideus* includes "a gathering together, a summing up, a uniting." This subsection, then, is a lift and gathering of results so far, in preparation for pre-heuristics to be obtained in the next section.

As pointed to in previous sections, 20th and 21st century advances in biochemistry are contributing to ongoing progress in chemical understanding of, for example, respiration, flight, and other pigeon activities. But, as already mentioned more than once, through self-attention one may notice and self-notice that in reaching chemical explanation of such activities, that is, in identifying metabolic profiles, results always are fragmentary. In particular, in each case one appeals to aggregates of biophysical; and biochemical boundary conditions. Through chemical law it is possible to explain diverse chemical capacities. But, as beginnings in self-attention reveal, neither physical law nor chemical law explain their own boundary conditions.

---

account of animal species will differentiate animals not by their organic but by their psychic differences. ... (T)he animal pertains to an explanatory genus beyond that of plant; that explanatory genus turns on sensibility; its specific differences are differences in sensibility; ..., possessing a degree of freedom that is limited, but not controlled, by underlying materials and outer circumstances" (*CWL3*, 290ff). Appealing to our own experience, within a future control of meaning, questions such as "What is it like to be a bat?" (Thomas Nagel, "What Is It Like to Be a Bat?" *The Philosophical Review*, vol. 83, No. 4, Duke University Press, (Oct., 1974): 435-450) will be able to contribute to scientifically significant description of animal experience.

If one is doing some of the biochemistry here (as one must, in order to make progress toward generalized empirical method), one may notice and self-notice that patterned sets of metabolic profiles provide patterned data too. So, might we not also ask after verifiable correlations between those metabolic profiles

$$(\textit{activity/function}) = <P(a): (p_i;c_j)_a, (p_i;c_j)_{(a,c)}, (p_i;c_j)_c > (\textit{activity/function})^{111}?$$

What are the empirical probabilities[112] of organism-wide profiles? Some may wonder, "Are these further enquiries really necessary?" Whatever the full heuristics will be, it is at least already evident and self-evident that without some such follow-up, there are patterned data and described activities that are not explained by chemistry and physics. Among other things, neither chemistry nor physics account for their own boundary conditions, let alone dynamically changing whole-organism metabolic profiles of boundary conditions.

Does this lead to logical problems? One way to begin to see that there needn't be a problem of logic is to note that inquiry into possible relations between metabolic profiles presupposes those profiles. Such inquiry does not deny the internal biophysical and biochemical coherence of profiles, or even of connectivities between profiles. On the contrary, the further inquiry presupposes already established profiles. But, the further investigation also is empirical. The dynamics of metabolic profiles reveal patterns in data not yet explained by the biophysics and biochemistry of individual profiles. Somewhat more precisely, it is found that vast aggregates of physical and chemical events occur regularly, in describable patterned combinations. And this calls us to further inquiry, and further understanding.

What, then, does *further* mean? We are barely touching on the core issue here. It is a major problem in contemporary science. One needs to enter into the empirical details of examples. Still, a few comments are possible. There is, then, no problem of logic. For, that further understanding will not directly involve the logic of chemical terms and relations. Instead, it will be new terms and new relations, verifiable in the

---

[111] See note 117.
[112] See Section 1.8.

patterned dynamic aggregates of actual whole organism profiles. This invites various key questions: What are those further terms and relations? What are corresponding probabilities of those new terms and relations? How many such layerings are there for the pigeon? And so on. These are all empirical questions for biologists studying pigeons. The point, here, is that even now empirical method calls us to such higher-order understanding of (to-be-discovered) relations between to-be-determined classes of organism-wide chemical pathways, without contradiction.[113] In the living pigeon, all orders are verified together - *"physics; and chemistry/but-not-merely physics; and chemistry/but-not-merely chemistry" – all deftly fitting like a dove.*

Symbolism is needed, which points to verifiable orders of properties of an organism, and at the same time points to a dynamic unity. A convenient symbol is the semi-colon ';'. Note that this also sheds light on the semi-colon symbolism throughout sections 5 and 6 above.[114]

---

[113] Analogous layerings of structures are familiar in mathematical development, and can be witnessed, for example, when one transitions from elementary arithmetic to elementary algebra; or elementary algebra to group theory. In generalized empirical method, one will seek precision in identifications and self-identifications. Series of increasingly higher-order logics also are familiar in contemporary mathematical logic, which includes Gödel's famous incompleteness theorems.

[114] Further studies are needed in order to investigate in detail how, in chemistry also, there is an analogous pairing: "physical-but-not-merely-physical," a pairing which becomes explanatory within a system that is a vast chemical matrix-mesh. But, is a special symbolism really needed for the various layerings? There is a recognized empirical problem, a multi-layerdness of things that are physical but not merely physical; chemical but not merely chemical; botanical but not merely botanical; and so on. At the very least, the semi-colon is a convenient way to point to a verifiable layeredness that also is a dynamic unity. The reader familiar with Lonergan's work may recall compact writings on *genus* and *species* as *explanatory* (*CWL3*, secs. 8.3-8.5). There, Lonergan provides similar symbolisms for things T_i, T_{ij} ..., and conjugates C_i, C_{ij}, ... In order to begin to fill out doctrinal pointers within an explanatory heuristics, we need to appeal to our detailed understandings in biophysics, biochemistry, ..., and make the effort to be luminous in details of specific and generic explanatory differentiations. Note, however, that working toward generalized empirical method, in basic position (*CWL3*, 413) we will need to grow in our heuristics so that we embrace not merely layerings of experiences and understandings, but also realities intended. In *Insight*, we see something of this (major) shift in Sec. 15.3, "Explanatory Genera and Species" (*CWL3*, 463-467). There, the

In contemporary biology and philosophy of biology, we do not yet find talk of "unity-identity-whole,"[115] of "not merely," or of "logical orders." But it *is* part of contemporary biology that, for the pigeon, there are known below-conscious functions; and known conscious functions. Might it seem, therefore, that there is nothing new here? For, in addition to the already known properties of the Autonomic Nervous System of the pigeon, it is well known that there are networks of afferent and efferent neurons of the Central and Peripheral Nervous Systems that are the anatomical-cellular basis for the flexible, dynamic and diversely combined functionings of the various special sense organs of the living pigeon. And, in contemporary biology, these results are known to be linked to the *psychology* of pigeons.

But, as pointed to briefly in Section 2.2, while there is progress, there is also ongoing confusion in philosophy of biology. And in traditional field biology, we find bifurcations and tunnelings within increasingly complex heterarchies of sub-disciplines, with few signs yet of an emerging shared heuristics.[116] And, the fact remains: it is the whole pigeon that flies, … , and rests, … , and breeds, … , *in situ*.

What is emerging here is the possibility of a verifiable heuristics for real layerings of properties of a living pigeon. Somewhat more precisely, these are modest intimations toward a (pre-) heuristics. Already, though, we can anticipate new questions: Are there verifiable classes of organism-wide *below-conscious-capacities* that, with respect to the most up-to-date understanding of below-conscious functions, are merely coincidental, yet occur regularly? In that case, there is patterned data inviting further explanation. And, by the same token, there is the need to work out a still further layering of higher-order terms and correlations, and probabilities. This further understanding would be of not just any chemical metabolic profiles $<P(a): (p_i;c_j)_a, (p_i;c_j)_{(a,c)}, (p_i;c_j)_c >$, but patterned aggregates of

---

symbolism is the same as in chapter 8 of *Insight*, but, chapter 15 of *Insight* speaks of *components of being*: central and conjugate potencies, forms and acts. Regarding organic forms, see 17.7.2, *"Organic Development"* (*CWL3*, 488-492). For more on the need of ;-symbolism, see Philip McShane, "Prehumous 2, Metagrams and Metaphysics," http://www.philipmcshane.ca/prehumous-02.pdf.

[115] See note 55.

[116] See note 5.

below-conscious profiles that are only partly accounted for, and otherwise non-systematic with respect to a prior layering of below-conscious properties.

As known in contemporary avian science, for the pigeon, there are at least four main layerings of properties: physical; chemical; below-conscious; conscious. And while in this chapter we have mainly looked in some detail at the pair "chemical; not merely chemical," that has been a help toward getting some intimation of the possibility of a full heuristics for the whole living pigeon, *in vivo* and *in situ*; for how layerings of properties function together in the pigeon: flexibly so, within activity profile ranges; with a remarkable *flexibility and mutual dependence* – all at once within-and-across ;-layers.[117]

---

[117] Eventually, mathematics for biology will need to be: verifiable in actual organisms; include structurings beyond the explanatory contexts of physics and chemistry; be topologically complex, that way allowing for partially overlapping classes of layered aggregates of events; and be functionally complex, allowing for dynamic mutual influences among layerings, as well as flexible groupings of classes of layered aggregates of events, and defensive schemes such as immune functions. In other words, the present exercise of attempting to develop a pre-heuristics for an adult pigeon can help one begin to glimpse something of the general heuristics outlined in *Insight*: In empirical method there is "an invitation to mathematicians to explore the possibility of setting up the series of deductive expansions that would do as much for other empirical sciences as has been done for physics" (*CWL3*, 339). In recent decades, there has been a growing openness in biology to the fundamental relevance of mathematical understanding. This is partly evidenced in ongoing mathematical work within systems biology. There is also a now globally recognized and established area of research called Mathematical Biology (see, for example, Society for Mathematical Biology, http://www.smb.org/index.shtml). Because of an increasing interplay between mathematics and biology, for some, there are now growing expectations that "Biology is the New Physics" (Philip Hunter, "Biology is the new physics," *European Molecular Biology Organization Reports* (*EMBO*), vol. 11, no. 5, May 1 (2010): http://embor.embopress.org/content/11/5/350) or that "Biology Is Mathematics' Next Physics, Only Better" (J.E. Cohen, "Mathematics Is Biology's Next Microscope, Only Better; Biology Is Mathematics' Next Physics, Only Better," *Public Library of Science, Biology* (*PLoS Biology*), 2(12), (2004): e439. A senior scholar in mathematical biology recently wrote: "Viewing the present trends in mathematical biology, I believe that the coming decade will demonstrate very clearly that mathematics is the future frontier of biology and biology is the future frontier of mathematics." (Avner Friedman, "What Is Mathematical Biology and How Useful Is It?" *Notices of the American Mathematical Society* (Aug. 2010): 857). However, this growing openness to mathematics in biology will

To conclude this section, think, now, of the pigeon, *in situ, in natura*. There are the bird's visual and olfactory capacities; feathers and nerves, an almost instantaneous sensitivity to changes in wind speed; capacities to consume oxygen and release carbon dioxide; functionalities of various neurotransmitter concentrations (e.g., serotonin); Autonomic and Central Nervous Systems; psychological affinities. Moment by moment, pulse by pulse, sight by sight, sound by sound, breath by breath, mood by mood, flight by flight, … , a pigeon reveals not only that it is multi-talented, but that it is deftly integral, able to shift through complexly integrally layered rhythms, not limited to phrasings of any one of physics, chemistry, below conscious or conscious. Instead, the pigeon is all of these at once. While having no fame for being a song bird, evidently, the pigeon is a (physical; chemical; below-conscious; and conscious) *aria maestro*.

---

need to be transformed within a future control of meaning. For, while there are various kinds of (sometimes multi-scale) micro- and macro- population dynamics, as already observed, many mathematical structures in contemporary mathematical biology are not verifiable in actual organisms. For examples, one may look to computational systems, computational biochemistry and computational biophysics. See Hunter's article (Philip Hunter, 2010). A similar list is in Friedman's review article (Friedman, 851-857). We may also look to Michael Reed's "Mathematical Biology," in the 2008 *Princeton Companion to Mathematics* (Michael C. Reed, "Mathematical Biology," Sec. VII.2 in *The Princeton Companion to Mathematics*, eds. Timothy Gowers, June Barrow-Green and Imre Leader (Princeton and Oxford: Princeton University Press, 2008), 837-848). A main purpose of Reed's article is "to illustrate, by selected examples (the) diversity and range of new mathematical questions that arise naturally in the biological sciences" (Reed, 837). Note, however, the emphasis on *mathematical* questions. The collections of examples given include applications of differential equations to fluid dynamics and multi-variable diffusion processes, genomics, the geometry and topology of macromolecules, biological fluid dynamics, phylogenetics and graph theory, and mathematics in medicine. In the case of neurobiology, the title of Reed's section 7 is: "What's wrong with Neurobiology?" "The short answer is that there is not (yet) enough theory" (Reed, 843). I note here that Woodger's 1960 work in terms of hierarchies may need to revisited (J. H. Woodger, "Biology and Physics," *The British Journal for the Philosophy of Science*, vol. XI, 42, Aug. (1960): 89 – 100). See also, Daniel J. Nicholson and Richard Gawne, "Rethinking Woodger's Legacy in the Philosophy of Biology," *Journal of the History of Biology*, 47 (2014):243–292.). Part of the challenge is that we also will need to include verifiable explanatory heuristics of development. Regarding development, see Chapter 3.

2.8 Fledgling Heuristics of an Adult Pigeon

Most of the details already have been discussed. Let us now bring these together within some suitable expression. At this stage, I now also bring in the more precise notation indicated in note 110. Consider, then, the formula C ($<P(a)>$))[a, (a,b), b] (p_i; c_j; b_k; z_l). The 'a' and 'b' are the various anatomical parts. Initially, we might well include triples of parenthetic terms (p_i; c_j; b_k; z_l)$_a$, (p_i; c_j; b_k; z_l) $_{(a,b)}$, (p_i; c_j; b_k; z_l)$_b$. But, symbolically, this is cumbersome. Instead, use a single (now inclusive) (p_i; c_j; b_k; z_l) with a multi-index [a, (a,b), b]. This simplifies notation somewhat; keeps the ;-layerings in view; and maintains [a, (a,b), b] as a symbolic pointer to an empirically verifiable anatomical basis of the various functionalities and connectivities. Finally, brace-brackets $<$, $>$ are for the verifiable dynamic unity, the *living span*.[118]

What about the *bold-face* letter 'C'? The pigeon is a Center of Capacities; 'C' is a symbol that Clasps, but also is *open* – a pointing, then, to the pigeon's open and dynamic integral Clasp of its ";-layered" Capacities. In fact, the etymology of 'capacity' includes "room enough to fit." The bold-face is in keeping with contemporary symbolisms, where bold-face usually means that there are many dimensions. Finally, conveniently for this chapter, 'C' also works as a reminder of the pigeon's family name, *Columbidae*, which, as we know, is but one of the pigeon's many Capacities.

2.9 Looking to Leave the Nest

As mentioned at the beginning of the chapter, sections 3 to 8 partly are inspired by Lonergan's two dense doctrinal paragraphs of *Insight*, 489-90. So far, a focus has been on the first of these, the paragraph which begins: "Study of an organism begins from the thing-for-us, from the organism as

---

[118] This is a temporary name. Avian science will find its own terminology. In the present context, I borrow the name 'span' from classical mathematics: In n-dimensional geometry, $<U, V, W, ...>$ is called the *span* of the vectors U, V, W, ..., that is, all possible vector sums and scalar multiples of U, V, W, ... Of course, I also think of 'wing-span' of the living bird.

exhibited to our senses."[119] A fledgling heuristics emerges for the adult pigeon. It is really a pre-heuristics, and is admittedly thin in various ways. The most basic way is that we have not entered into any depth in biological understanding. In particular, the below-conscious and conscious capacities have been more alluded to, than described in any up-to-date scientific context. Nevertheless, a pre-heuristics is obtained, a pre-heuristics that is both empirically grounded and open to development.

Questions arise, now, that call us to heuristics of what is not yet accounted for in this chapter. Pigeons don't start their lives as adult birds. *Columbidae* grows from being a blind embryo, to a vulnerable squab, to an inelegant fledgling, eventually to being an impressive air master. In other words, there is avian *development*.

There are also many kinds of things besides pigeons. Some are said to be *alive*, like the vast multitudes of animals, plants, and microorganisms. There are also viruses, "at the edge of life."[120] At the same time, other things are said to be *not alive*, like stones, physical and chemical compounds, and elementary particles. Moreover, as contemporary environmental science reveals, all things both living and not living are in mutual dependence. And there is more. The totality, the concrete intelligibility of Space and Time that Lonergan called "emergent probability"[121] is evidently dynamic. New species of organism emerge, old species become extinct; stars are born, galaxies form and others collapse; and we too, who embrace galaxies and more, are part of this dynamic totality, individually and historically.

Evidently, the curve of generalized empirical method steepens considerably. The paragraphs from *Insight* 489ff provide helpful pointers. Where contemporary biologists verify properties and laws of developing organisms, Lonergan invites us to major development within a basic position. In the second of those two paragraphs of *Insight* 489-90, Lonergan refers to "laws," "schemes of recurrence," and "capacities-to-

---

[119] *CWL3*, 489.

[120] E. P. Rybicki, "The classification of organisms at the edge of life, or problems with virus systematics," *South African Journal of Science*, 86 (1990):182–186. Viruses do not have their own metabolism, and require a host cell to replicate. See also Chapter 4.

[121] *CWL3*, secs. 4.2.4 and 5.5.

perform." He then goes on to write about "transposition to the thing-itself."[122] He then writes that such a transposition reveals sets of "conjugate forms,"[123] a "higher system as integrator,"[124] ... , and a "higher system as operator."[125] The second paragraph of *Insight* 489 continues:

> As integrator, this set is related (1) to inspected organs as the set of functions grasped by the physiologist in the sensible data, (2) to the physical, chemical, and cytological manifold as the conjugates implicitly defined by the correlations that account for addition regularities in the otherwise coincidental manifold, and (3) to immanent and transient activities of the organism in its environment as the ground of the flexible circle of ranges of schemes of recurrence. However, the organism grows and develops. Its higher system at any stage of development not only is an integrator but also an operator, that is, it so integrates the underlying manifold as to call forth, by the principles of correspondence and emergence, its own replacement by a more specific and effective integrator.[126]

To move toward the larger context, the next chapter will look to the problem of organic development. This presents us with new and difficult problems. For, whatever else the new generalized empirical method is to be, it will be empirical. In particular, generalized empirical method does not admit conceptual models or imaginable representations that cannot be verified. In order to reach a pre-heuristics of organic development, it will necessary to enter some ways into biological understanding of development. But, this book does not presuppose expertise in biology. And so we will need to continue working with elementary examples. Taking that course, it is possible to make some progress in at least getting helpful first impressions of realities intended. Note that as mentioned in the Introduction of this book, after Chapter 3 on organic development, Chapter 4 goes on to invite precision about issues that arise in the first three chapters. For example, some help will be taken from both St. Thomas Aquinas and modern science in our reach for glimmerings of what Lonergan was pointing to when speaking of "transposition to the thing itself," "conjugate form," "operator" and "integrator."

---

[122] *CWL3*, 489. See also Chapter 4.

[123] *CWL3*, 489.

[124] *CWL3*, 489.

[125] *CWL3*, 489.

[126] *CWL3*, 489-90.

Chapter 3

Growing to Flight Mastery: The Whole Storeyed Story

Abstract: There are three main contexts for this chapter. There is biology and its progress in understanding development; there is philosophy of biological development; and, there is the ongoing invitation to generalized empirical method. As throughout the present text, the main purpose here is not to solve specific problems within biology and philosophy of biology. Solutions will emerge within a future generalized empirical method. The more immediate purpose here is to help further bring out the need of an improved empirical method; and to glimpse something of the possibility of verifiable heuristics of biological development. Special attention will be given to avian development, one of the most studied in 20th century embryology. The outline of the chapter is as follows: Section 1 recalls a few aspects of views from philosophy of development in biology. Section 2 comments briefly on Lonergan's heuristics of development. Section 3 reviews elementary descriptions of development. Section 4 invites attention to details of avian embryonic development, known in contemporary avian science. Section 5 is on the problem of obtaining verifiable explanatory heuristics for avian development. Section 6 provides a few comments regarding biological development and genetic method.

3.1 Philosophies of Development in Biology

Since ancient times, there have been efforts to understand the development of organisms. Recall, for example, Aristotle's well-known investigations in zoology, his studies of chick embryos, and more.[1] In modern science, development, as such, became an object of study in the late 19th century. Philosophy of biology emerged as an established specialization in the 20th century.[2] Since the discovery of DNA by Watson and Cricks,[3] one of the central issues in philosophy of developmental biology has been stated as follows:

> If nearly all cells of an organism contain the same genes, what system of controls determines the patterned differentiation and movement of cells that is required to make animal bodies? It is hard to see how the specifications that determine the structure of the body and its orderly construction can be contained in the genes alone. But if the genes alone do not contain the "program" specifying the structure and properties of an organism (in recent terms: if the nucleotide sequence of the DNA of a fertilized egg does not, by itself, contain the information necessary to specify the structure of the organism that will result from the normal development of that egg), we need to (1) rethink the prevalent genetic determinism and our use of the metaphor of a genetic program, and (2) work out in detail the system of controls that make ontogenesis as reliable as it is. … Such issues have been actively entertained for more than a century.[4]

Burian goes on to say:

> It is now clear that various controls that determine the cascades of events that produce segments (and also some particular organs …) are preserved across enormous evolutionary distances. … This finding is critical for the argument that multilevel causation is involved in both development and evolution. For one thing, within the life of a single organism, the same genetic material is used at different stages of ontogeny and in different tissues in quite different ways. (In this chapter (12) I explore some of the consequences this has for how we think about genes. It

---

[1] James Lennox, trans., *Aristotle: On the Parts of Animals I-IV*. Gloucestershire: Clarendon Press, 2002. See also, James Lennox, "Aristotle's Biology," *The Stanford Encyclopedia of Philosophy* (Spring 2014 Edition), Edward N. Zalta, ed., http://plato.stanford.edu/entries/aristotle-biology/.

[2] David L. Hull, "The History of the Philosophy of Biology," ch. 1 in Michael Ruse, ed., *The Oxford Handbook of Philosophy of Biology* (Oxford: Oxford University Press, 2008), 11 – 33.

[3] Historical context is given in, Leslie A. Pray, "Discovery of DNA structure and function: Watson and Crick," *Nature Education* 1 (1) (2008): 100. http://www.nature.com/scitable.

[4] Richard Burian, "Part IV, Development," in *The Epistemology of Development, Evolution, and Genetics, Selected Essays, Cambridge Studies in Philosophy and Biology* (Cambridge: Cambridge University Press, 2005), 179-180.

appears that many genes are composed of functional subunits that, on an evolutionary scale, are swapped from one gene to another.) ... . And it forces a deep reconsideration of the ways in which animals are constructed.[5]

For molecular and developmental biology, a convenient summary of main views, past and present, is provided by Griffiths.[6] A few samplings follow:

The current literature in the philosophy of molecular and development biology has grown out of (earlier) discussions under the influence of twenty years of rapid and exciting growth of empirical knowledge. Philosophers have examined the concepts of genetic information and genetic program, competing definitions of the gene itself, and competing accounts over the relationship between development and evolutions.

...

According to the classical account of theory reduction, one theory reduces to another when the laws and generalization of the first theory can be deduced from those of the second theory with the help of bridge principles relating the vocabularies of the two theories.[7]

One of the more recent views being promoted is *biocomplexity and self-organization.*

Stuart Kauffman's (1993) simulations of networks of 'genetic' elements suggested that basic biological phenomena such as autocatalytic cycles required for the origin of life or the array of cell-types required for the emergence of multi-cellular life are highly probable outcomes of random variation in complex chemical or, later, genetic networks.[8]

In "Current Status of Problems," [9] Griffiths also provides summaries and literature references for recent *genetic information* theories, *genetic program* theories, *developmental systems* theories, controversies about the

---

[5] Richard Burian, "Part IV, Development," 181.

[6] Paul Griffiths, "Molecular and Developmental Biology," ch. 12 in *The Blackwell Guide to the Philosophy of Science*, eds. Peter Machamer and Michael Silberstein (Malden, Massachusetts and Oxford: Blackwell, 2002), 252-271.

[7] Griffiths, 252.

[8] Griffiths, 259.

[9] Griffiths, 259-266.

gene concept, as well as views of development that include the effects of environment, and evolution.

> Despite the ubiquity of talk of genetic information in molecular and developmental biology, the predominant view in recent philosophical work on this topic has been that 'genetic information' and 'genetic program' have a precise meaning only in the context of the relationship between DNA sequence, RNA sequence, and protein structure. In their broader applications, these ideas are merely picturesque ways to talk about correlations and causation.

>   The obvious way to explicate information talk in biology is via information theory.

> ...

> Alexander Rosenberg has defended the view that the study of development is the study of how the embryo is 'computed' from the genes and proteins contained in the egg cell. ... Keller has rejected this interpretation of the science, arguing that gene activation in the developing embryo is precisely not like the unfolding of a stored program, but instead like a distributed computing, in which processes are reliably executed by local interactions in networks of simple elements.

> ...

> DST (developmental systems theory) is an alternative to the relationship between genes and other factors in development. ... DST argues for a thoroughgoing epigenetic[10] account of development. ... biological form must be reconstructed in each generation by interaction between physical causes. ... The term 'developmental system' refers to the system of physical resources that interact to produce the life-cycle of a particular evolving lineage.

> ...

> Schaeffner has argued that most work in the molecular developmental biology conforms to DST (developmental systems theory) strictures about distributed control of development and the context sensitivity of genetic and other causes.[11]

---

[10] Epigenetic: a: of, relating to, or produced by the chain of developmental processes ... that lead from genotype to phenotype after the initial action of the genes; b: relating to, being, or involving changes in gene function that do not involve changes in DNA sequence (http://www.merriam-webster.com/dictionary/epigenetic).

[11] Griffiths 260-263.

These quotations are but a small sampling of Griffiths' detailed chapter. Still, they are somewhat representative. Griffiths outlines a variety of views, and observes, also, that there is an increasing dissatisfaction with earlier genetic program models. As a result, a more recent trend is to suppose models with more complex programs, parallel computing, distributed computing and other system-wide networks.

If we look to the work of Mary West-Eberhard, though, we find a different view. First, she recalls a few main features of the alleged genetic program: The *genetic program* is a

> predominant metaphor for the organization of development and evolution, which describes development as programmed by the genes and reprogrammable by genetic change during evolution. An entire set of instructions for development is contained in the genes. ... The computer analogy allows for input from the environment. But, ..., the rules and outcomes of interactions are completely defined by the genes. Development begins with genetic instructions, and evolved programming begins with genetic change.[12]

West-Eberhard then goes on to point out various difficulties with the metaphor:

> The genetic program metaphor does not suggest the possibility that environmental elements are partly or entirely responsible for the development (or nondevelopment) of a phenotypic trait. It is a metaphor that leads to thinking in terms of genetic directives.
>
> ...
>
> The complete-instruction metaphors are particularly problematic because they reinforce the misconception held by many that the genome is a complete set of instructions for making an organism. This image has little resemblance to the decentralized way that responsive structure is made during ontogeny, with the form and function of tissues and organs depending importantly on their circumstances and uses as they are being formed.[13]

Jason Scott Robert also expresses concern about the gene-focused approach to development:

---

[12] Mary Jane West-Eberhard, *Developmental Plasticity and Evolution* (Oxford: Oxford University Press, 2003), 14.
[13] West-Eberhard, 15.

Taking development into account is not the same as taking development seriously. To take development seriously is not to hide behind metaphors of the magical powers of genes –they 'instruct' or 'program' the future organism. To take development seriously is rather to explore in detail the processes and mechanisms of differentiation, morphogenesis, and growth, and the actual (not ideologically or technologically inflated) roles of genes in these organismal activities. Despite the existence of what has come to be known as the "interactionist consensus," according to which everyone agrees that both genes and environments 'interact' in the generation (and explanation) of organismal traits, my claim is that those swept up in genomania have nonetheless failed to take development seriously.[14]

Where concerns are expressed by West-Eberhard, Scott Robert and others, Griffiths suggests that:

· the predominant view in recent philosophical work on this topic has been that 'genetic information' and 'genetic program' have a precise meaning only in the context of the relationship between DNA sequence, RNA sequence, and protein structure. In their broader applications, these ideas are merely picturesque ways to talk about correlations and causation. The obvious way to explicate information talk in biology is via information theory.[15]

There are, though, evident problems in the methods that give rise to the various metaphors. Recall, from the work of Chapter 2, that a biochemical program is neither verifiable in particular organisms nor compatible with biological practice. But, that incompatibility only increases when one attempts to bring the same idea to the study of development of, say, a particular pigeon over its entire life-cycle. Whether it is an older genetic program or a more recent model of systems of distributed programs and system-wide networks, part of the assertion is there are programs between DNA, RNA and protein structures. But, as we have already seen, that hypothesis is neither experimentally observed nor, more crucially, are they biochemically verified.

Just as for systems theory generally, it is not that there is no supporting data at all for developmental systems theory. There is ongoing progress in discovering and verifying wonderfully complex biochemistry of intra-cellular and inter-cellular biochemical pathways. And, as also mentioned

---

[14] Jason Robert Scott, *Embryology, Epigenesis and Evolution. Taking Development Seriously* (Cambridge: Cambridge University Press, 2004), xiii.

[15] See Griffiths, 260.

in Chapter 2, contemporary systems biology is regularly making advances in biochemistry for medicine.[16] Again, though, how are these advances obtained and verified? Laboratory methods require complex combinations of sequences of biochemical reactions and analyses, and depend also on statistical methods for what typically are multiple samplings, large numbers of cells and numerous tissue samples. What, then, of a particular amoeba, or pigeon, or dog or cat? One team of systems biologists comments on a few aspects of the problem:

(I)t is important to bear in mind that the graphs generated by these tools may reflect a compilation of molecular events that occur throughout the lifetime of the cell or organism, depending on the method that was used to delineate the network, and that not all interactions are necessarily direct or have a regulatory consequence.[17]

That systems models of development are remote to living organisms is made explicit in the following:

First, we wish to make clear that our definition of systems biology is primarily conceptual rather than one dependent on technology, or data mass. For us, systems developmental biology is the means of achieving a causal explanation of development, which begins with the genomic regulatory information and extends vertically to all the different levels of biological organization directly affected by the developmental genomic control system. The explanation must encompass both the spatial allocation of gene expression and its temporal sequence. This is fundamentally conceptual as well as an experimental discovery problem in information processing and regulatory logic, on a system-wide scale. Second, by

---

[16] Also referenced in Chapter 2, systems biology continues to have "profound implications for cancer research" (Pau Creixell et al., "Navigating cancer network attractors for tumor-specific therapy." *Nature Biotechnology* 30, (2012): 842-848). For example, see K. A. Janes, et al. "A systems model of signaling identifies a molecular basis set for cytokine-induced apoptosis," *Science* 310, (2005):1646–1653. Similar results are ongoing. See, for example, Joseph Loscalzo and Albert-Laszlo Barabasi, "Systems Biology and the Future of Medicine," *Wiley's Interdisciplinary Reviews: Systems Biology and Medicine*, Vol. 3 (2011): 619-627. Another recent review article, on systems biology in biochemistry for medicine, is Miguel Angel Medina, "Systems biology for molecular life sciences and its impact in biomedicine," *Cellular and Molecular Life Sciences*, 70 (2013):1035–1053. doi 10.1007/s00018-012-1109-z.

[17] Martha L. Bulyk and A. J. Marian Walhout, "Gene Regulatory Networks," ch. 4 in *Handbook of Systems Biology, Concepts and Insights*, eds. Marian Walhout, Marc Vidal and Job Dekker (Amsterdam: Academic Press, 2013), 65-88.

alluding briefly to the deep roots of this conceptual problem, we can see that it was to a large extent formulated independently of the gigantic mass of current knowledge of molecular, biochemical, and cell biological detail. Indeed, the subject of this chapter is just how the 'material configuration' of the control system that Wilson deduced actually causes the process which, from then until now, has been termed 'development.'[18]

In systems biology, there is ongoing progress in being able to work out (intervals for) attainable boundary conditions in biochemical pathways of organisms. And in philosophy of biological development, critical issues are being revealed. However, is there not also something evidently problematic in contemporary methods, when genetic programs and information networks *per se* are primarily conceptual, and not verifiable in actual organisms? As Griffiths observes, however, the

debate over the role of information concepts in biology is in full swing at the present and likely to continue. The renewed contact between the philosophies of evolutionary and developmental biology is also likely to occupy many writers for some time to come.[19]

Surely, something more is needed.

3.2   Lonergan's Heuristics of Development

As can be seen from the previous section, "the notion of development is peculiarly subject to the distorting influence of counterpositions."[20] Lonergan's leads on development are in sections 15.6-15.7 of *Insight*.

How, then, is a concrete instance of development to be investigated?

One has to follow the lead of the successful scientists, the physicists and the chemists, but one has to imitate them not slavishly but intelligently.[21]

---

[18] Isabelle S Peter and Eric H. Davidson, "Transcriptional Network Logic: The Systems Biology of Development," ch. 11 in in *Handbook of Systems Biology, Concepts and Insights*, eds. Marian Walhout, Marc Vidal and Job Dekker (Amsterdam: Academic Press, 2013), 211-228. See p. 213.

[19] Griffiths, "Future Work," 267,

[20] *Insight, CWL3*, 476.

[21] *CWL3*, 488ff, secs. 15.6-15.7.

He goes on to a heuristic definition:

a development may be defined as a flexible, linked sequence of dynamic and increasingly differentiated higher integrations that meet the tension of successively transformed underlying manifolds through successive applications of the principles of correspondence and emergence.

...

However, lest this prove a mere jumble of words, let us add to the illustrations of parts of the definition a few illustrations of the whole. [22]

Part of the difficulty, however, is that Lonergan's "few illustrations"[23] take up only a few pages of *Insight*. And, they are not so much illustrative for the reader as they are touching on extensive work-to-be done by the community. Of course, merely becoming familiar with Lonergan's dense expressions is not development toward explanatory heuristics of biological development. And, if one is to work at the level of the times, data on development is to be obtained through relatively up-to-date scientific inquiries and understandings of instances of biological development.

The next three sections of this chapter, then, are to encourage the emergence of that future possibility. The present effort necessarily will be superficial, but will be to help stimulate a budding from the visceral arch that is present-day avian science. For, if we follow up within a growing self-attention, we can begin to get a few inklings of the possibility of being luminous in a heuristics of biological development.

3.3　Avian Development

In Chapter 2, part of the focus was on the family called *columbidae*. It remains convenient to continue drawing on results about the pigeon or dove. Among other things, they have been the object of extensive and ongoing modern research about biological development.[24]

---

[22] *CWL3*, 479

[23] *CWL3*, 479-484.

[24] "The chicken embryo has been used as a model system for embryology and develop-mental biology for more than 2 millennia" (L. A. Cogburn et al., "Functional Genomics of

To begin, let's look to a nest of mourning doves (*zenaida macroura columbidae*). Figure 3.1 provides four relevant images. Some similar example within the reader's experience will do.

Hatching and growth			
 Egg in nest	 Nesting in progress	 Squabs	 A juvenile

Fig. 3.1. Hatching and growth of mourning dove (*zenaida macroura columbidae*).[25]

If we say 'mourning dove,' we are in that nest of human capacities to describe which includes the capacity to 'name.'[26] But, is not one not getting at something special when one not only says 'mourning dove,' but also says also that 'the mourning dove is growing'? Is there not also an

---

the Chicken—A Model Organism," *Poultry Science*, vol. 86, issue 10 (2007): 2059). See also, C. D. Stern, "The chick embryo—Past, present and future as a model system in developmental biology," *Mechanisms of Development*, 121 (2004):1011–1013. C. D. Stern, "The chick: A great model system becomes even greater," *Developmental Cell*, vol.8, issue 1 (Jan. 2005): 9–17.

[25] http://en.wikipedia.org/wiki/Mourning_Dove. Open Access. Creative Commons.

[26] See note 55 of Chapter 2.

understanding that, while something is happening, all along the way it is the same dove? In the context of the present chapter, this is an exercise that calls for elementary self-attention. And, if one continues to investigate the growth of mourning doves, eventually one is able to describe a regular pattern of change that includes: egg, chick, squab, juvenile, first flight, and beyond.

Preliminary description of dynamics of 'naming,' as well as that grasp of change in the organism will both be core to the problem of becoming luminous about one's understanding of development. However, generalized empirical method will reach beyond elementary description. For now, however, within the context of the present book, we can attend to some of what scientific understanding has been learning about what, for birds, is a special stage in avian development.

3.4 Wing Buds

An adult bird develops from a single haploid ovum and sperm. ... The development of the embryo involves cell division and cell differentiation. From an initial undifferentiated cell, specialised cells arise at different times and different positions within the developing embryo. ... During the earlier part of the (20th) century, the avian embryo was one of the most studied species by embryologists, and the morphological stages of development were described in detail (Lillie, 1919). ... As development proceeds, cells in different parts of the embryo are clearly destined to become particular structures in the adult. When the avian embryo has reached the blastula[27] stage, it is possible to draw a fate map, since the destination of cells according to their position has been determined.[28]

The embryo grows on. Hamburger and Hamilton are known for their studies of the anatomy of avian embryonic development.[29] They identified

---

[27] blastula: from Greek (blastos), meaning "sprout", hollow sphere of cells, or blastomeres, produced during the development of an embryo by repeated cleavage of a fertilized egg. The cells of the blastula form an epithelial (covering) layer, called the blastoderm, enclosing a fluid-filled cavity, the blastocoel. See, for example, *Encyclopaedia Brittanica*, http://www.britannica.com.

[28] Lewis Stevens, *Avian Biochemistry and Molecular Biology* (Cambridge: Cambridge University Press: 1996), 195.

[29] Viktor Hamburger and Howard L. Hamilton, "A Series of Normal Stages in the Development of the Chick Embryo," *Developmental Dynamics* 195 (1992): 231- 272. (Reprinted from *Journal of Morphology*, vol. 88, no. 1, (1951): 49–92.) Their stage 1 does

46 main stages of growth in embryonic development of domestic fowl, after an egg is laid. The stages

are well defined, and ... are used as reference points in the study of development.[30]

Their paper contains detailed descriptions and a representative sequence of photo plates. While there are differences for other avian species,

(t)oday, most descriptions of avian embryonic development refer to the normal stages identified by Hamburger and Hamilton,[31]

obtained in 1951. For example, for *columbidae*, there are 42 such stages.

A few of the descriptions from the Hamburger and Hamilton paper are included, and will be referred to below.

Stage 1. Pre-Streak: Prior to the appearance of the primitive streak. An "embryonic shield" may be visible, due to the accumulation of cells toward the posterior half of the blastoderm.

...

Beyond stage 14 the number of somites becomes increasingly difficult to determine with accuracy. This is due in part to the dispersal of the mesoderm of the anteriormost somites, and, in later stages, to the curvature of the tail. Total somite-counts given for the following stages are typical, but sufficiently variable so as not to be diagnostic. For these reasons, the limb-buds, visceral arches, and other externally visible structures are used as identifying criteria from stage 15 onward.

Stage 15. (ca. 50-55 hrs.). 1. Lateral body-folds extend to anterior end of wing-level (somites 15-17). 2. Limb-primordia: prospective limb-areas flat, not yet demarcated. Inconspicuous condensation of mesoderm in wing-level. 3. Somites: 24-27. 4. Amnion extends to somites 7-14. 5. Flexures and rotation. Cranial flexure: axes of

---

not go right back to the zygote, but is when "an 'embryonic shield may be visible, due to the accumulation of cells toward the posterior half of the blastoderm" (Hamburger and Hamilton (1992): 54).

[30] Stevens, 195.

[31] Robert E. Ricklefs and J. Mathias Stark, "Embryonic Growth and Development," ch. 2 in *Avian Growth and Development: Evolution Within the Altricial-precocial Spectrum* (Oxford: Oxford University Press, 1998), 39-40. Approximately 31 avian species are represented in their Table 2.3, p. 40. In that table, the Hamburger-Hamilton embryonic stages range from 38 to 46, depending on species.

forebrain and hindbrain form an acute angle. The ventral contours of forebrain and hindbrain are nearly parallel. Cervical flexure a broad curve. The trunk is distinct. Rotation extends to somites 11 to 13. 6. Visceral arches: Visceral arch 3 and cleft 3 are distinct. The latter is shorter than cleft 2 and usually oval in shape. 7. Eye: Optic cup is completely formed; double contour distinct in region of iris.

Stage 16. (ca. 51-56 hrs.). Lateral body-folds extend to somites 17-20, between levels of wings and legs. Limbs. Wing is lifted off blastoderm by infolding of lateral body-fold. It is represented by a thickened ridge. Primordium of leg is still flat; represented by a condensation of mesoderm. 3. Somites: 26-28. 4. Amnion extends to somites 10-18. 5. Flexures and rotation: All flexures are more accentuated. 6. Tail-bud a short, straight cone, delimited from blastoderm. 7. Visceral arches: Third cleft still oval in shape. 8. Forebrain lengthened; constrictions between brain-parts are deepened. Epiphysis indistinct or not yet formed.

...

Stage 45. (ca. 19-20 days). Beak: Length is no longer diagnostic; in fact, the beak is usually shorter than in stage 44, due to a loss (by sloughing off) of its entire peridermal covering. As a consequence, the beak is now shiny all over and more blunt at its tip. Both labial grooves have disappeared with the periderm. 2. Third toe: Average length is essentially unchanged from that of stage 44, except in those breeds with a longer period of incubation (21 days) and a heavier build of body. For these latter, length of third toe = ca. 21.4 i 0.8 mm. 3. Extra-embryonic membranes: Yolk-sac is half-enclosed in body-cavity. Chorio-allantoic membrane contains less blood and is "sticky" in the living embryo.

Stage 46. Newly-hatched chick (20-21 days).[32]

Perhaps not surprisingly, the emergence and development of wing buds has been one of the most studied aggregate of events in avian science. In the 15th stage of the Hamburger-Hamilton scale, there are no wing buds present in the embryo. In the 16th stage, however, wing buds have emerged. Let's look at this part of avian embryonic development in a little more detail.

Limb bud development is initiated by the release of specialised mesenchymal cells from the somatic layer of the lateral plate, followed by their migration laterally and accumulation under the epithelial tissue. ... The positional information, along the

---

[32] See note 29.

anterior-posterior limb axis, originates from the *zone of polarizing activity*, which is the likely source of the morphogen retinoic acid.

...

Retinoic acid is formed by the oxidation of retinol, and both retinoic acid and the related metabolite, 3,4-didehydroretinoic acid, are present in the normal limb buds of chick embryos (Scholfield, Rowe & Brickell, 1992), the latter at substantially higher concentration (Mavilio, 1993). ... Although retinoic acid is assumed to have a role in pattern formation, it may act synergistically with other morphogens, or there are several steps to the process.[33]

In keeping with our notation from Chapter 2, C can represent the whole bird. To include development, for the present discussion we can appeal to the Hamburger and Hamilton avian embryo development scale. So, for the 15th stage, write C_{15}; and for the 16th stage, C_{16}. Note that this is drawing on empirical results.

How are the Hamburger-Hamilton stages distinguished? In the quotation above, Stevens mentions that the Hamburger-Hamilton scale "is well defined."[34] There is an invariance in the anatomical differentiations observed by Hamburger and Hamilton.[35] But, Stevens' book focuses on more recent developments in avian *biochemistry*, with a 12th chapter on the "Molecular genetics of avian development."[36] What has been found is that there are patterned aggregates of cellular, biochemical and biophysical events in the Hamburger-Hamilton 16th stage, that are not present in the Hamburger-Hamilton 15th stage. Of course, similar observations apply to other transitions between Hamburger-Hamilton stages. But, for now, let's continue with our focus on the transition from the 15th to the 16th stage.

Recall that Chapter 2 leads up to a heuristics for a multi-layered (physical; chemical; below conscious; conscious) *adult* bird, or in the more

---

[33] Stevens, 200-202. Italics in source text.

[34] Note 30.

[35] From early on, immune functions emerge. See note 53. In some cases development becomes abnormal. Within limits, an embryo sometimes can re-acquire normal growth trajectory. See note 54.

[36] Lewis, Stevens, ch. 12, 195–211.

precise terminology, (physical; chemical; botanical; zoological).[37] A question here is: Do we need the full four-fold multi-layered heuristics for all embryonic stages 1 to 46? Even before the Hamburger-Hamilton's first stage, there are changes in the embryo that occur before the formation of the "embryonic shield" (*Stage 1*).[38] And, if we go back as far as the original zygote, there are hundreds of metabolic pathways, vast and complexly patterned aggregates of biochemical events, merely coincidental relative to the systematics of chemistry. Moreover, as the zygote moves toward division, there are further remarkable changes in the biochemistry of the cell. In other words, from the beginning, it is evident that the embryo is chemical, but not merely chemical. And so, with the new symbolism, the chick zygote is represented by an at least triple-layering (p_i; c_j; b_k; _), and for the 15th stage we would write $C_{15}(p_i; c_j; b_k; \_)$.

Notice that, for the moment, a space '_' has been left in the z-position. Is the pigeon embryo conscious? In a 2007 study, it was reported that

> electroencephalographic (EEG) activity, which evolves subsequently, shows that states of sleep-like unconsciousness … predominate in chicks until after hatching.[39]

---

[37] Recall symbolism from Chapter 2: (physical; chemical; botanical; zoological) is abbreviated to (pi; cj; bk; zl). Note that subscripts 'i', 'j', 'k', 'l' are crucial. They heuristically point to verifiable complex heterarchic-like orderings known and being discovered through ongoing avian biophysics; avian biochemistry; and so on. The semicolon ';' is a key feature of the symbolism, pointing to a verifiable (flexible and dynamic) multi-layeredness that is physical, but not merely physical; chemical but not merely chemical; and so on.

[38] This is not a complaint about the important work of Hamburger and Hamilton. And, anatomy of earlier development of the embryo was not easily investigated with apparatus available at the time.

[39] D.J. Mellor and T.J. Diesch, "Birth and hatching: Key events in the onset of awareness in the lamb and chick," *New Zealand Veterinary Journal*, 55(2), (2007): 51-60. More recently: "After hatching, the chick usually takes at least 2 h, and possibly longer, before it exhibits sustained behavioural and EEG evidence of conscious wakefulness (Mellor and Dietsch, 2007). Note, however, that its EEG is relatively mature at hatching. Pigeon chicks, on the other hand (Table 10.1), apparently exhibit EEG silence during the first three days and there is a slow evolution of the EEG to more mature forms between days 6 and 14. This slower developmental pattern has been attributed to greater cerebral cortical immaturity in the hatchling pigeon than in the domestic chick (Ellington and Rose, 1970).

However, the chick is moving through its stages of embryonic development, moving toward consciousness. And, scientific understandding of avian development is ongoing. As Mellor, Patterson-Kane and Stafford observed,

(t)his appears to merit further investigation.[40]

To allow for ongoing progress, we can leave the z_k as a place holder for whatever biologists discover about consciousness in stages of avian development. The subscripts ('15', '16', and so on) can then refer to Hamburger-Hamilton stages of development, as they emerge.

Let us now bring this together somewhat. Appealing to the Hamburger-Hamilton stages, we need a full multi-layered heuristics for both the 15th and 16th stages. In the new symbolism, stages 15 and 16 can be written as $C_{15}(p_i; c_j; b_k; z_l)$ and $C_{16}(p_i; c_j; b_k; z_l)$. As biology makes progress, understanding of stages and transition dynamics between stages also grows. Additional stages or refinements may be discovered, or perhaps the Hamburger-Hamilton stages will be replaced. There may be new precision in matching certain later stages of neural development with corresponding stages of animal consciousness. But, for now, let us keep our focus on *differences* between the Hamilton-Hamburger stages, provisionally, C_{15} and C_{16}. In $C_{16}(p_i; c_j; b_k; z_l)$, there are new specialized cells, anatomically localized, retinoic acid and other morphogen events, none of which are present in $C_{15}(p_i; c_j; b_k; z_l)$. And, there are many other known differences between C_{15} and C_{16}. But, even one such difference is enough to bring out something of the puzzle that we call development.

There is some understanding of the multi-layered $(p_i; c_j; b_k; z_l)$ dynamics verifiable in the 15th stage. According to avian science, there is a 15th stage in which the $(p_i; c_j; b_k; z_l)$ layerings verifiably function

---

The range of neurological maturity at hatching suggested by these few observations in avian species (see also Rogers, 1995) may therefore parallel that seen in mammalian young at birth (marsupials excluded)(Table 10.1). This appears to merit further investigation" (David Mellor, Emily Patterson-Kane, Kevin J. Stafford, *The Sciences of Animal Welfare* (Oxford: Wiley-Blackwell, 2009), 179).

[40] David Mellor, Emily Patterson-Kane, Kevin J. Stafford, (2009): 179. See note 39.

together, integrally and flexibly, within and across layerings, within verifiable statistical ranges of boundary conditions. An integrality is understood. And, it is (self-) evident that understanding integrality is something different from understanding, for example, electron transport processes in mitochondrial membranes (in the TCA cycle); or a collection of biochemical reaction equations; or a capacity of the organism to secure nutrition from available resources and eliminate foreign organisms. The integrality is an understood and known reality of the embryonic bird. The young bird is an *integrator*.[41]

How is the *integrator* studied?[42] Implicitly, biology already gives us a lot to go on. There are various stages of development. The anatomical changes and (p_i; c_j; b_k; z_l) multi-layered sophistications of each stage are becoming increasingly known. At this time, emerging details of each of the 46 (or for the pigeon, 42) Hamburger-Hamilton stages are filling biology journals of the world; and new results are published regularly.

However, as the organism continues to develop from the 15th stage, it transitions toward a 16th stage. According to present-day biology and biochemistry, neither the transition nor the next stage are accounted for within the anatomy and the (p_i; c_j; b_k; z_l) dynamics of C_{15}. Relative to C_{15}, something new is happening. Verifiably, the entire (p_i; c_j; b_k; z_l) multi-layered embryo is, in fact, shifting ground. But, the change is not arbitrary: a pigeon embryo normally develops into a pigeon hatchling, and a pigeon hatchling into an adult pigeon. Able to succeed under diverse circumstances, an embryo moves from a 15th stage to a 16th stage, in verifiably specific and highly nuanced ways. Recall, for example, the anatomical and biochemical details of the transition described in Stevens' book:

Limb bud development is initiated ...[43]

---

[41] The possibility of identifying *metaphysical equivalents* will be raised in Chapter 4.

[42] See note 44.

[43] See note 33. Much of the rest of Stevens' dense Chapter 12 points to further biochemical detail (as well as needed research) on development of the avian nervous system, groups of cell adhesion molecules and other proteins.

In mathematics, an *operator* is a transformation between spaces of functions. Of course, avian science is not mathematics. But,

there is an extremely important point to the mathematical analogy.[44]

The avian embryo has a 15th stage. And, it is also verifiable that the embryo works its way toward a 16th stage. In other words, in addition to being an *integrator*, the bird is also an *organic operator*[45]: it verifiably transforms its own 15th stage of (p_i; c_j; b_k; z_l) functionings to a new 16th stage of (p_i; c_j; b_k; z_l) functionings.

How is the operator studied? ... The difficulty in studying the operator lies in the complexity of its data.[46]

But, as centuries of embryology and studies in biology show, progress is being made.[47] And results to-date reveal that the avian embryo transitions from one stage to a next stage in specific ways, details of which are gradually being discovered within the developing science that is avian embryology.

3.5 All Stages of Avian Development

The focus in the previous section was on just two stages of embryonic development, a 15th stage immediately prior to the emergence of wing buds; and transition to a 16th stage in which wing buds are present. The Hamburger-Hamilton *Stage 1* starts after an egg is laid.[48] However, growth of the embryo begins well before that, and in fact begins with a zygote. And, at the end of the Hamilton-Hamburger scale (46th stage for the domestic fowl, 42nd for *columbidae*[49]), the *hatchling* emerges from the

---

[44] *CWL3*, 490.

[45] As for *integrator*, here too, the problem of determining metaphysical equivalents will be raised in Chapter 4.

[46] *CWL3*, 491.

[47] See, for example, note 43.

[48] See the description of *Stage 1* in the extract of the Hamburger-Hamilton paper, note 32.

[49] See note 31.

shell, with still further stages of development to follow which include, for example, squab, juvenile, and adult.[50]

If we now lift elementary results from Chapter 2 to the present context, while still remote, the possibility of a verifiable heuristics of development begins to emerge. A heuristics will need to include all verifiable "well defined"[51] stages. If we want indexing to line up with the results of Hamburger and Hamilton, then, temporarily, the index '0' can stand for (every stage of) development that occurs prior to an egg being laid, from zygote to just prior to the Hamburger-Hamilton *Stage 1*. But, it is progress in biology that will determine the extent to which refinements of *Stage 0* are needed and explanatory. (In fact, Hamburger-Hamilton already identified three such stages prior to an egg being laid.) Next, we can continue the indexing through all intermediate stages of the Hamburger-Hamilton sequence, and then beyond, to a *Stage a*, 'a' for 'adult'. Putting all of this together, the avian organism $C(p_i; c_j; b_k; z_l)$ develops through a verifiable time-ordered sequence of stages of growth:

$$C_0(p_i; c_j; b_k; z_l) \rightarrow C_1(p_i; c_j; b_k; z_l) \rightarrow \ldots \rightarrow C_a(p_i; c_j; b_k; z_l).$$

Notation can be improved somewhat. For one thing, the indexing '1, 2, 3, …' and the arrows '$\rightarrow$' might suggest discrete steps and a rigid path of development. But, growth processes and transition dynamics continue throughout development. And, while stages are "well defined,"[52] there is also tremendous variation. Even in mitosis of a single zygote, formations emerge that are merely transitional. In the more mature embryo, some *fate maps* are known. And, basic defensive capacities begin to emerge within the first three days of fertilization.[53] It is, nevertheless, normal that many

---

[50] There is also aging, a gradual loss in the organism's capacities-to-perform.

[51] Note 30.

[52] Note 30.

[53] Banislav D. Janković, Zoran Knežović, Ljilja Kojić and Vera Nikolić, "Pinneal Gland and Immune System, Immune Functions in the Chick Embryo Pinealectomized at 96 hours of Incubation," *Annals of the New York Academy of Sciences*, 719 (1994): 398–409. (Article first published online: 8 JUN 2007.) B. D. Janković et al., "Immunological capacity of the chicken embryo. I. Relationship between the maturation of lymphoid tissues and the occurrence of cell-mediated immunity in the developing chicken embryo." *Immunology*, 29 (3) (Sept. 1975): 497–508. B. D. Janković, K. Isaković, B. M. Marković,

cellular formations emerge that serve only transitional functions. In such cases, cellular formations are either relocated, change or are sloughed off. There is also the possibility of abnormal development. But, in some cases at least, a bird can recover normal development.[54] Note also that a main feature of avian development is illustrated by the Hamburger-Hamilton avian stages of development: each stage $C_{s+1}(p_i; c_j; b_k; z_l)$ is not fully accounted for within prior stages C_s, C_{s-1}, ..., C_0. A provisional heuristics needs to allow for all of these subtleties, flexibilities, capacities and alternative development routes, some of which already are partially understood; as well as what might yet be discoverable.

So, instead of indexing discrete steps, let's just write stage 's,' and that way allow for future revisions, refinements and discoveries beyond the Hamburger-Hamilton scale. A symbolism for stage 's' then becomes $C_s(p_i; c_j; b_k; z_l)$. Avian development not only is multi-layered ($p_i; c_j; b_k; z_l$), but is emergent, from prior stages, a kind of organic sum. A familiar symbol for 'sum' is the symbol '$\int$' from integral calculus. As is well known, the symbol is a stylized 'S' invented by Leibniz, 'S' for *summa* (Latin), or 'sum.' Here, though, the sum is not of a mathematical function. Instead, it is a *summa* of ongoing ($p_i; c_j; b_k; z_l$) organic development: $\int C_t(p_i; c_j; b_k; z_l)\, dt$.[55]

In biochemistry, the name 'genetic' is for the nucleotide chemical structuring of DNA and RNA. But, the word 'genetic' also has an older

---

and M. Rajcević, "Immunological capacity of the chicken embryo. II. Humoral immune responses in embryos and young chickens bursectomized and sham-bursectomized at 52--64 h of incubation." *Immunology*. 32(5) (may, 1977): 689–699.

[54] The literature in avian science on recovery of normal development goes back decades. Two examples are: Christopher T. Goode et al., "Visual Influences on the Development and Recovery of the Vestibuloocular Reflex in the Chicken." *Journal of Neurophysiology*, vol. 85, issue 3 (2001): 1119-1128. J. K. Kovach, "Development of pecking behavior in chicks: Recovery after deprivation." *Journal of Comparative and Physiological Psychology*, vol. 68, no. 4 (Aug 1969): 516-523.

[55] At all stages of development, there are dynamics of both integrator and operator. Each stage $C_s(p_i; c_j; b_k; z_l)$ will be defined through explanatory terms and relations proper to that stage of development. As in Chapter 2 on the adult dove, for each stage there will be multi-layered statistical results of multi-layered boundary conditions. But, the developing organism continues to shift ground. So, there also will be nuanced layerings of statistics of nuanced layerings of change that reveal the propensities and flexibilities of the operator.

and broader meaning, namely, *origin* and *generation*. With that broader meaning, let us call $C_t(p_i; c_j; b_k; z_l)$ dt a *genetic sequence*.[56] Note that in this context, a genetic sequence includes biochemical genetic sub-structurings[57] of DNA and RNA.

3.6 Development in Biology

Where does the work of the previous section get us? The pre-heuristics developed is empirical, provisional, verifiable, and allows for $(p_i; c_j; b_k; z_l)$ properties yet to be discovered in avian science. If developed within an emerging generalized empirical method, such a heuristics will lift what is already known[58] into new control of meaning. It also will help biology and philosophy of biology escape the net of networks of systems biology. For, while systems biology has been producing positive results, in its essential hypotheses it has proven to be non-verifiable.

---

[56] *CWL3*, 486. See also note 59.

[57] The meaning of "*sub*-structurings" is to be determined within a developing $(p_i; c_j; b_k; z_l)$ heuristics already discussed. Verifiably, an organism is chemical-but-not-merely-chemical. As biochemistry has been learning, a particular DNA sequence of base pairs limits ranges of possible higher capacities by which an organism can survive in an environment.

[58] See, for example, J. Matthias Starck et al, eds., *Avian Growth and Development: Evolution Within the Altricial-precocial Spectrum*. Oxford: Oxford University Press, 1998. The book brings together an extensive body of advances in developmental avian science. In the fifteenth chapter, however, three biologists reflect on what so far are three main classes of model of avian growth. "(These) (m)odels applied to growth (in birds) serve several purposes. For the most part, (these) models have been used to describe growth empirically, ..., characterizing growth by parameter values of simple equations, ... (e.g. logistic equations). ... A second class of models has been developed to help organize thinking about physiological mechanisms ... A third class of models analyzes functional aspects in an evolutionary context" (Marek Konarzewski, Sebaastian A.L.M. Kooijman and Robert E. Ricklefs, "Models for Avian Growth and Development," ch. 15 in, J. Matthias Starck et al., 340). "(T)he testing of (these) models is often very difficult because most of the parameters cannot be isolated and varied experimentally" (Marek Konarzewski et al, 361). The need for an adequate heuristics is being recognized within biology, and points to future work that will be possible within generalized empirical method. See also notes in sec. 2.7, above.

We are, though, far from implementation of an improved empirical method. The pre-heuristics of organic development outlined above are merely preliminary and descriptive searchings, mere wing buds to a future control of meaning in avian science. To end this chapter, I recall Lonergan's precise doctrinal statement regarding genetic method. The quotation reveals something of his control of meaning. But, it also points to what will be possible, at some stage, before too long:

> Genetic method is concerned with sequences in which correlations and regularities change. Accordingly, the principal object of genetic method is to master the sequence itself, to understand the development, and thereby to proceed from the correlations and regularities of one stage to those of the next.[59]

---

[59] *CWL3*, 486.

Chapter 4

Biological Entities

Abstract: The chapter begins by recalling Schrödinger's 1944 Dublin lecture series on physics, chemistry and biology, and his leads on the question 'What is Life?' Section 2 points briefly to a tradition of thought that has been follow-up to Schrödinger's lectures. That way we get some indication of the extensive development that has occurred, but we also get still further evidence on the pressing need of development in method. Section 3 recalls a description given by Thomas Aquinas, of knowing when something is 'alive.' Section 4 recalls Lonergan's definitions of conjugate and central potency, form and act, a major shift in heuristics called for in the transition from *Insight*,[1] Chapter 8 to *Insight*, Chapter 15. Section 5 appeals to some of what is known about chemotaxis of one-celled organisms. With some work, this helps bring out the possibility of future developments in heuristics of organic matter. Section 6 outlines further questions. These regard: viruses; organic *potentia activa*; negentropy; information and communication theories; and recent efforts in mathematical modeling of biological processes. Section 7 points to a future control of meaning.

---

[1] Bernard Lonergan, *Insight: A Study of Human Understanding*, vol. 3 in Frederick E. Crowe and Robert M. Doran, eds. *Collected Works of Bernard Lonergan* (Toronto: University of Toronto Press, 1992).

4.1 Schrödinger's Question: What is Life?

In the famous 1944 Trinity College Dublin lectures,[2] speaking about physics, chemistry and organisms, Schrödinger asked the question: 'What is Life?'

> The large and very much discussed question is: How can the events in space and time which take place within the spatial boundary of an organism be accounted for by physics and chemistry?[3]

In chapter six of his book, he gave various suggestions toward an answer:

> What is the characteristic feature of life? When is a piece of matter said to be alive? When it goes on 'doing something,'   moving, exchanging material with its environment, and so forth, and that for a much longer period than we would expect of an inanimate piece of matter to 'keep it going' under similar circumstances.[4]

The same subsection of Schrödinger's chapter 6 ends with:

> What an organism feeds upon is negative entropy. Or, to put it less paradoxically, the essential thing in metabolism is that the organism succeeds in freeing itself from all of the entropy it cannot help producing while alive.[5]

There are several reasons why the present chapter begins with Schrödinger's question. Schrödinger's lectures brought together various fundamental issues. There have been significant advances in molecular biology that were, in part at least, inspired by Schrödinger's work. There is also a body of philosophic literature that has followed up on the richness of his thought on physics, chemistry and biology.[6] While it has been more

---

[2] Erwin Schrödinger, *What is Life? The Physical Aspect of the Living Cell, with, Mind and Matter, and Autobiographical Notes.* Cambridge: Cambridge University Press, 1994. There have been several editions. See also, http://whatislife.stanford.edu/LoCo_files/What-is-Life.pdf.

[3] Schrödinger, *What is Life?*, chs. 1 and 3.

[4] *What is Life?*, ch. 6, 69.

[5] *What is Life?*, ch. 6, 71.

[6] A discussion of the relevance of his ideas can be found in, Hans Ulrich Gumbrecht at al., *What is Life? The Intellectual Pertinence of Erwin Schrödinger.* Stanford: Stanford University Press, 2011.

than 70 years since Schrödinger's lectures, the question 'What is Life?' "has remained frustratingly out of our grasp for decades, and perhaps centuries."[7] As Schrödinger indicated, and as the ongoing scientific and philosophic traditions reveal, the question is important to physics, chemistry, biology, philosophy of biology and philosophy of science. In particular, it has become evident that progress toward answering the question 'What is Life?' will need to include a heuristics of energy and entropy. There are now established areas of research relating energy, entropy and thermodynamics to fundamental questions in ecology and evolution.[8] Lonergan, too, points to the importance of answering basic questions about energy in chemistry, biology, and the universe.[9]

Of course, this chapter is not intended to provide an answer to Schrödinger's question 'What is Life?' However, with an eye on method, reflecting on the question here can help us toward anticipating certain aspects of that progress.

4.2 The Living Question

How can the events in space and time which take place within the spatial boundary of an organism be accounted for by physics and chemistry?[10]

Each organism has a physics and a chemistry, and since Schrödinger's lectures in 1944, there has been an ongoing effort in the community to reach for increasingly comprehensive views. A helpful synopsis of main views at this time is given by Deplazes-Zemp and Biller-Andorno.[11] In

---

[7] Niles Leman, "What is Life? How Chemistry Becomes Biology," Review of Addy Pross, *What is Life? How Chemistry Becomes Biology.* Oxford: Oxford University Press, 2012/2014, in *Trends in Evolutionary Biology*, vol. 5, no. 1, 2013.

[8] See, for example, Sven Erik Jorgensen and Yuri M. Svirezhev, *Towards a Thermodynamic Theory for Ecological Systems.* Oxford: Pergamon (Elsevier), 2004.

[9] *CWL3*, 469.

[10] See note 3.

[11] Anna Deplazes-Zemp and Nikola Biller-Andorno, "Explaining life. Synthetic biology and non-scientific understanding of life," *EMBO reports (European Molecular Biology Organization)*, vol. 13, no. 11 (2012): 959-963. Contributions to "furthering the understanding of life" (Ann Deplazes-Zemp, "The Conception of Life in Synthetic Biology," *Science and Engineering Ethics*, 18 (2012): 758) also come from Synthetic

their report to the European Molecular Biology Organization (*EMBO*), views were

> divided roughly into three main groups: scientific, philosophic and religious. However, there are not three clearly separable and uniform categories; they overlap with one another and each category comprises subcategories that may not have much in common with others. In this sense (they) have chosen pronounced positions that represent this wide range, rather than a representative selection of the most prevalent views.[12]

A similar observation was made by Macklem and Seely:

> a clear idea of what life means, remains elusive, and there is no universally accepted definition.[13]

Macklem and Seely do, though, offer their own description:

> Life is a self-contained, self-regulating, self-organizing, self-reproducing, inter-connected, open thermodynamic network of component parts which performs work, existing in a complex regime which combines stability and adaptability in the phase transition between order and chaos, as a plant, animal, fungus, or microbe.[14]

As in recent views on organic development,[15] the program metaphor is found throughout the contemporary literature on the question 'What is Life?' For example, there is Koshland's list of 'Seven Pillars of Life,' the first pillar of which he calls 'Program':

> By program I mean an organized plan that describes both the ingredients themselves and the kinetics of the interactions among ingredients as the living system persists through time. For the living systems we observe on Earth, this program is implemented by the DNA that encodes the genes of Earth's organisms and that is replicated from generation to generation, with small changes but always with the

---

Biology, "a set of different scientific and technological disciplines, which share the objective to design and produce new life forms" (Ann Deplazes-Zemp, 757). Although, the emphasis in this area is less on attempting to explain what life is, and is more on producing new life forms.

[12] Deplazes-Zemp and Biller-Andorno, 959.

[13] Peter T. Macklem and Andrew Seely, "Towards a Definition of Life," *Perspectives in Biology and Medicine*, vol. 53, no. 3 (Summer, 2013): 330-340.

[14] Macklem and Seely, 331.

[15] See Chapter 3.

overall plan intact. The genes in turn encode for chemicals—the proteins, nucleic acids, etc.—that carry out the reactions in living systems. It is in the DNA that the program is summarized and maintained for life on Earth.[16]

A consequence of views about alleged DNA coding and programs is that information theory now also features in leading hypotheses about organic life. For an example, we can look to De Loof's article,[17] in which it is suggested that there are eight "traits of the living state,"[18] the first of which is:

Communication is so essential to the living state at all its different levels of organization, from the cell organelle to the population that in my opinion life could be defined as follows:

$$\text{Life} = \int_{\text{cell organelle}}^{\text{population}} \text{communication}$$

Death ensues when the ability to communicate at the highest level of organization of the biological system under consideration is irreversibly lost.[19]

The remaining seven traits given by De Loof speak to various aspects of communication. The eighth and last in the list is:

The definition given under 1. or, more practically, "life: the ability to communicate" may bring more unity in all the different meanings which the different disciplines studying "life" intuitively attribute to their study object.[20]

A recently emerging area of research, *biosemiotics*, seeks to combine information theory with semantics. As stated by one scholar in the area,

a biological process keeps its order through the use of information. For defining order, it is proposed that an orderly object can be produced by a construction (e.g.,

---

[16] Daniel. E. Jr. Koshland, "The Seven Pillars of Life," *Science*, 295 (2002): 2215-2216. The 'Seven Pillars' in Koshland's list are: Program, Improvisation, Compartmentalization, Energy, Regeneration, Adaptability, and Seclusion.

[17] Arnold De Loof, "Schrödinger 50 Years Ago: What is Life? 'The Ability to Communicate', A Plausible Reply?" *International Journal of Biochemistry*, vol. 25, no. 12 (1993): 1715-1721.

[18] De Loof, 1715.

[19] De Loof, 1715.

[20] De Loof, 1715.

the copy of a template) using available data within some given context. In other words, replicating an orderly object does not bring new information into its context. Order in this meaning appears as specific to the living world, at variance with the inanimate world which is basically disorderly. A better understanding of what separates the living world from the inanimate world results: the use of information is the distinguishing feature which defines their border. Any living thing contains a symbolic information, referred to as its genome, inscribed into DNA molecules.[21]

The mission statement of the *Springer* journal *Biosemiotics* sheds further light on the area: The journal's

coverage spans a range of disciplines, bridging biology, philosophy, linguistics and the communication sciences. Conceived in the insight that the genetic code is a language as old as life itself, and grounded in the study of signs, of communication and of information in organisms, biosemiotics is evolving today toward the challenge of naturalizing not only biological information but also biological meaning, in the belief that signs and codes are fundamental components of the living world.[22]

However, as mentioned in the *EMBO* report,[23] there is not yet consensus on the main question. A group in systems biology suggests that:

a less ambitious, but more pragmatic question could be, what does Life require? Biologists agree on at least four fundamental requirements, among which are … *chemistry*. … *genes* to encode and 'reproduce' biomolecules. … *cells* provide the fundamental medium in which biological processes take place. … *evolution* by natural selection. … In the last decade, novel biological questions have surfaced, or resurfaced, pointing to *systems* as a fifth fundamental requirement for Life. Although conceptual, systems turn out to be as crucial to biology as chemistry, genes, cells or evolution.[24]

In the conclusions of their article, Carvunis et al write:

It will also be essential to achieve systematic integration of three-dimensional structural data, whether derived experimentally or by computational modeling.

---

[21] Gérard Battail, "An Answer to Schrödinger's '*What is Life?*'" *Biosemiotics*, 4 (2011): 55-56.

[22] *Biosemiotics*, Online journal, http://link.springer.com/journal/12304.

[23] See note 12.

[24] Anne-Ruxandra Carvunis et al., "Interactome Networks," ch. 3 in, A. J. Marian Walhout, Marc Vidal and Job Dekker, eds., *Handbook of Systems Biology, Concepts and Insights* (Amsterdam: Academic Press, 2013), 45.

Eventually, three-dimensional mapping of the sequence variations found in populations and their association with traits may allow the almost seamless reconstruction of genotype-phenotype relationships through edgetic modeling of protein-protein interactions.[25]

Other recent work promotes a more exclusively chemical-systems theory approach. The work of Addy Pross, for example, focuses on dynamic chemical stabilities (DKS) of living organisms.[26] From this perspective

the essential difference between the (chemical and biological) is just that the chemical phase (abiogenesis) is the low complexity segment of the single process, while the biological phase (Darwinian evolution) is the high complexity segment (Pross 2011, 2012).[27]

In a more recent article, Pross writes:

(A) biological process keeps its order through the use of information. For defining order, it is proposed that an orderly object can be produced by a construction (e.g., the copy of a template) using available data within some given context. In other words, replicating an orderly object does not bring new information into its context. Order in this meaning appears as specific to the living world, at variance with the inanimate world which is basically disorderly. A better understanding of what separates the living world from the inanimate world results: the use of information is the distinguishing feature which defines their border. Any living thing contains a symbolic information, referred to as its genome, inscribed into DNA molecules.[28]

Pross also suggests that:

recent developments in the relatively new area of systems chemistry have been showing that the reactivity patterns of simple replicating systems may assist in the building of conceptual bridges between the physicochemical (inanimate) and biological (animate) worlds.[29]

---

[25] Anne-Ruxandra Carvunis et al., 58.

[26] Addy Pross, *What is Life? How Chemistry becomes Biology*. Oxford: Oxford University Press, 2014.

[27] Addy Pross, "How Does Biology Emerge From Chemistry? Conference Report," *Origins of Life and Evolution of Biospheres*, 42 (2012):435.

[28] Addy Pross and Robert Pascal, "The origin of life: what we know, what we can know and what we will never know," *Open Biology*, 3 (2013): 120190, Royal Society, 90.

[29] Addy Pross, "How Does Biology Emerge From Chemistry? Conference Report," 433.

And, it has been found that:

a key element in that effort has been the ability to specify and characterize a new kind of stability– dynamic kinetic stability (DKS), one that pertains to replicating systems, whether chemical or biological.[30]

There is ongoing progress in 21st century field biology and laboratory biology. For example, Pross and others who investigate (verifiable) chemical kinetics of organisms have been making fundamental advances in determining chemical potentialities and chemical limitations of organisms.[31] Such work is pushing the chemical-envelope magnificently. But, as revealed in samples from the literature given above, and as may be increasingly evident to the reader, philosophic reflection on the nature of that progress struggles within various ongoing misconceptions. For example, as in theories of development,[32] here too main views attempt to explain organic life in terms of programs and networks, and distinguish chemistry, botany and zoology merely by degree of: chemical order, chemical phase transitions and chemical stability. But, as already observed in chapters two and three, there are no data available on programs *per se*, or networks *per se*.[33] Neither is there data on "symbolic information, ..., inscribed into DNA molecules,"[34] nor data to reveal that a

living thing contains a symbolic information, referred to as its genome, inscribed into DNA molecules.'[35]

Progress in biochemistry shows that diagrammatic representations of proteins, folded proteins (conformations) and protein-protein interactions

---

[30] Addy Pross, "How Does Biology Emerge From Chemistry? Conference Report," 433. See also, Addy Pross, "Dynamic Kinetic Stability (DKS) as a Conceptual Bridge Linking Chemistry to Biology," *Current Organic Chemistry*, vol. 17, no. 16 (2013): 1702-1703.

[31] See, for example, Addy Pross, *Theoretical and physical principles of organic reactivity*. New York: John Wiley & Sons, 1995.

[32] See Chapter 3.

[33] The word 'network' is a name for collections of linked pathways and cycles. See next paragraph, which recalls the empirical significance of biophysical and biochemical pathways and cycles.

[34] Gérard Battail, 55-56.

[35] Pross and Pascal, 120190.

can be helpful. But, recall how such representations are obtained. One-, two-, and three-dimensional representations are diagrams or material constructs (nowadays, usually computer generated) obtained from numerical simulation techniques and statistical analyses of large numerical data sets obtained from many other numerical simulations, and from large numbers of other physical and chemical experimental results, amalgamated from different laboratories from around the world, past and present.[36]

This is not to suggest that there is no data on individual molecules, chemical networks, or pathways. And, imaging techniques regularly are being improved.[37] But, even when shapes of individual molecular forms

---

[36] The literature on molecular simulation is large. See, for example, Luca Monticelli, Emppu Salonen, eds., *Biomolecular Simulations, Methods and Protocols, Methods in Molecular Biology*, vol. 924 of *Methods in Molecular Biology*, New York: Humana Press, Spring, 2013. "Much of the impressive progress in biomolecular simulations has been simply due to more powerful computers" (Luca Monticelli, Emppu Salonen, v). "Molecular simulations have become fundamental tools in science and engineering. They include a broad range of methodologies such as Monte Carlo, Brownian dynamics, lattice dynamics, and molecular dynamics (MD). Applications of molecular simulations to increasingly complex systems have been stimulated by tremendous progress in the development of computer hardware and software including new and faster algorithms and precise methods" Perla B. Balbuena and Jorge M. Seminario, eds., *Molecular Dynamics From Classical to Quantum Methods*, vol. 7 of *Theoretical and Computational Chemistry*. Amsterdam: Elsevier, 1999. "The calculation of atomic interactions or contacts from an MD (Molecular Dynamics) trajectory is computationally demanding and the work required grows exponentially with the size of the simulation system" (Rudesh D. Toofanny et al., "Methodology article, Implementation of 3D spatial indexing and compression in a large-scale molecular dynamics simulation database for rapid atomic contact detection." *BMC Bioinformatics* 2011, vol. 12: 334 (2011):1471-2105.

[37] See, for example, Martin Guthold et al., "Direct Observation of One-Dimensional Diffusion and Transcription by *Escherichia coli* RNA Polymerase," *Biophysical Journal*, vol. 77, issue 4 (October 1999): 2284–2294. "Heretofore, nearly all investigations have used a large population of molecules because few techniques were available to study single transcription complexes. Hence, important properties of the individual molecules may have been lost in the time and population averages involved in bulk studies" (Martin Guthold et al., 2285). In that work, "scanning force microscope (SFM) imaging in liquid was used to investigate the dynamics of single, nonspecific, and specific RNAP-DNA complexes in real time" (Martin Guthold et al., 2285). In these experiments, the presence of individual complex molecules is revealed by thin edgings in laboratory photo plates. See also, Ji Hoon

are obtained through new imaging techniques, an image is still just an image. And, what of data on chemical pathways? Chemical pathways are now standard in biochemistry. From biochemistry labs of the world, there are data on thousands of biophysical and biochemical cycles and pathways. But, as already observed in chapter two, such results are extremely remote to particular organisms, and are fragmentary.

Nevertheless, it has been suggested that it may be possible to construct

conceptual bridges between the physicochemical (inanimate) and biological (animate) worlds.[38]

What, though, about an actual organism? As already brought out somewhat in the previous chapter, in a living pigeon say, we find patterns in combinations of whole-organism aggregates of biophysical and biochemical events that, from the point of view of chemistry and physics, are merely coincidental.

Rehmann-Sutter, a philosopher of biology, takes a different approach,[39] that is, he does not attempt to construct a conceptual bridge. Instead, he reflects on

what we mean when we say that a ... cell is alive.[40]

He does not try to differentiate living and non-living as merely apparent differences. He describes what he calls "aliveness":

Kim and Ronald G. Larson, "Single-molecule analysis of 1D diffusion and transcription elongation of T7 RNA polymerase along individual stretched DNA molecules," *Nucleic Acids Research*, vol. 35, issue 11 (June, 2007): 3848–3858: "Figure 1, Experimental setup for visualizing DNA–T7 RNAP interactions" (Ji Hoon Kim and Ronald G. Larson, 2007). Ralf Jungmann et al., "Multiplexed 3D cellular super-resolution imaging with DNA-PAINT and Exchange-PAINT," *Nature Methods*, issue 11 (2014): 313-318. See also, Nicole A. Becker and L. James Maher, "High-resolution mapping of architectural DNA binding protein facilitation of a DNA repression loop in Escherichia coli," *Proceedings of the National Academy of Sciences (PNAS)*, vol. 112, no. 23 (June 9, 2015): 7177–7182.
[38] See note 29.
[39] Christoph Rehmann-Sutter, "How Do We See that Something is Living? Synthetic Creatures and Phenomenology of Perception," *Worldviews*, 17 (2013): 10-25.
[40] Rehmann-Sutter, 10.

a special way of being that is somehow qualitatively different from the being of a non-living thing.[41]

Rehmann-Sutter asks:

What must be given clearly and how should it be given, in order that the object is perceived as living?[42]

Rehmann-Sutter's paper is descriptive and does not reach into the contemporary context of biochemistry and modern science. But, he observes how one can be startled if what one thought was a dry leaf in a web turns out to be a spider that suddenly moves about. He also comments on the experience of having a friendly pet dog. A first part of his two-part answer to the question is that living things

seem to have their own way of being because they own 'their own center' in themselves.[43]

Marvelous advances are being made in 21st century biophysics and biochemistry. At the same time, we also speak of living things and non-living things. We say that a dog is alive; and that a stone is not. What, then, is Life? The above samplings from the literature provide only a glimpse into recent work on the question. Still, it is enough to reveal that there is ongoing confusion. On the positive side, however, we also find an increasing pressure in the biosciences and philosophy of biology for an improved control of meaning.

4.3 Describing Our Knowing About Living Things

One may be amazed by a bird in flight; be familiar with the comings and goings of one's pet cat or dog, or, as Rehmann-Sutter found, be startled by

---

[41] Rehmann-Sutter, 11.

[42] Rehmann-Sutter, 18.

[43] Rehmann-Sutter, 11.

the movements of a spider. Perhaps, like van Leeuwenhoek and Hooke, one may be fascinated by the many living things not usually visible to the naked eye.[44] That there are some things that we call 'living' and other things 'not-living' is not in question.

How, though, can we approach the problem in a way so that eventually we will be able to handle the problem at the level of the times? Rehmann-Sutter asks that we not exclude descriptions and normal experience. Thomas Aquinas also provides his description of when we know that something is 'alive.'[45] In the 13th century, however, biochemistry and modern biology were still centuries away. Nevertheless, Aquinas was a genius. And, we may take some help from his reflections. Like Rehmann-Sutter, Aquinas' reflections also invite one to attend to experience. The precision of Aquinas, though, challenges the reader to self-attention. And, that exercise can help us in our both in our search for clues about what organic life is and in making progress in empirical method.

On the question "Do all natural things have life?"[46] Aquinas answers the following:

> *I answer that*, We can gather to what things life belongs, and to what it does not, from such things as manifestly possess life. Now life manifestly belongs to animals, for it said in *De Vegetabilbus*. i [*De Plantis* i, 1] that in animals life is manifest. We must, therefore, distinguish living from lifeless things, by comparing them to that by reason of which animals are said to live: and this it is in which life is manifested

---

[44] "The existence of microscopic organisms was discovered during the period 1665-83 by two Fellows of The Royal Society, Robert Hooke and Antoni van Leeuwenhoek. In *Micrographia* (1665), Hooke presented the first published depiction of a microrganism, the microfungus *Mucor*. Later, Leeuwenhoek observed and described microscopic protozoa and bacteria. These important revelations were made possible by the ingenuity of Hooke and Leeuwenhoek in fabricating and using simple microscopes that magnified objects from about 25-fold to 250-fold" (Howard Gest, "The Discovery of Microorganisms by Robert Hooke and Antoni van Leeuwenhoek, Fellows of the Royal Society," *Notes and Records, The Royal Society Journal of the History of Science*, 58(2), (2004): 187).

[45] Thomas Aquinas, *Summa Theoligica*, First Part, Question 18, Article 1. See, for example, http://www.newadvent.org/summa/1018.htm. For a more recent scholar's translation, see Thomas Gilby, O. P., trans., *Summa Theologiae*, St. Thomas Aquinas, New Edition. Cambridge: Cambridge University Press, 2006.

[46] Thomas Aquinas, *Summa Theologica*, First Part, Q18, Art. 1, "Do All Natural Things Have Life?"

first and remains last. We say then that an animal begins to live when it begins to move of itself: and as long as such movement appears in it, so long as it is considered to be alive. When it no longer has any movement of itself, but is only moved by another power, then its life is said to fail, and the animal to be dead. Whereby it is clear that those things are properly called living that move themselves by some kind of movement, whether it be movement properly so called, as the act of an imperfect being, i.e. of a thing in potentiality, is called movement; or movement in a more general sense, as when said of the act of a perfect thing, as understanding and feeling are called movement. Accordingly all things are said to be alive that determine themselves to movement or operation of any kind: whereas those things that cannot by their nature do so, cannot be called living, unless by a similitude.[47]

The exercise here is to read the answer given by Aquinas, line by line, with some example of one's own. Aquinas moves on to subtleties about 'movement' that are not yet a topic in this book. But, how does his description work for you? In practice at least, do we not often determine which things have life and which do not? And, does not the description compare well your own experience?:

We say then that an animal begins to live when it begins to move of itself: and as long as such movement appears in it, so long as it is considered to be alive. [48]

4.4 Potencies, Forms and Acts of Living Things

The previous section invites preliminary exercises in self-attention. There is the possibility of making elementary progress in describing something of how "we gather to what things life belongs."[49] And, even though it is one's own gathering, are we not usually getting at something more than just one's own gathering? If you gather that a living pigeon is on the balcony, are you not gathering that that it *is so* that a pigeon *is* on the balcony? And, if you gather that a pigeon *flies* to the balcony, are you not gathering that it *is so* that a pigeon *flies* to the balcony? Even while there is the "enveloping world of sense,"[50] we also intend what *is*.

Taking help from Aristotle, Thomas Aquinas was able to work out a

---

[47] Thomas Aquinas, *Summa Theologica*, First Part, Q18, Art. 1, "Do All Natural Things Have Life?"

[48] *Summa Theologica*, First Part, Q18, Art. 1.

[49] *Summa Theologica*, First Part, Q18, Art. 1.

[50] *CWL3*, 559.

descriptive but precise heuristics of beings in terms of potency, form and act.[51] But, if the genius Thomas was at home in such a heuristics, science was in its infancy. Both Mendeleev and Krebs were centuries away, and in particular, there was no question yet of *How Chemistry becomes Biology*.[52] Whatever Thomas meant is a problem for distant future interpretations. For the present context, there is our present challenge, the need to move at least a little way toward some improved control of meaning about our various gatherings in modern science.

In *Insight* we find the following (ultimately helpful, but as we will see shortly, only apparently easy-to-read) descriptions of *potency*, *form* and *act*. These will be familiar to readers of Lonergan's work but, for the convenience of all readers, I include the full statements:

'Potency' denotes the component of proportionate being to be known in a fully explanatory knowledge by an intellectually patterned experience of the empirical residue.

'Form' denotes the component of proportionate being to known, not by understanding the name of things, nor by understanding their relations to us, but by understanding them fully in their relations to one another.

'Act' denotes the component of proportionate being to known by uttering the virtually unconditioned yes of reasonable judgment.[53]

Later in the same chapter fifteen of *Insight*, Lonergan draws attention to different kinds of forms:

(T)here is a fundamental heuristic structure that leads to the determination of conjugates, that is, of terms defined implicitly by their empirically verified and explanatory relations. Such terms as related are known by understanding, and so they are forms. Let us name them conjugate forms.

...

Further, the heuristic structure that leads to knowledge of conjugate forms necessitates another structure that leads to knowledge of central forms. For one reaches explanatory conjugates by considering data as similar to other data; but the

---

[51] See, for example, Thomas Aquinas, *De Principiis Naturae*.
[52] Note 26.
[53] *CWL3*, 457.

data which are similar also are concrete and individual; and as concrete and individual, they are understood inasmuch as one grasps in them a concrete and intelligible unity, identity, whole.[54]

These descriptions can seem so reasonable. This will be so especially if one already has some experience with describing one's "cognitional process"[55]: experience, understanding and judgment. The diagram on page 299 of *Insight* may be familiar, and so too might the two diagrams in Appendix A of *Phenomenology and Logic, The Boston College Lecture Series on Mathematical Logic and Existentialism*.[56] One might find oneself comfortable with the analogy: experience is to understanding is to judgment as potency is to form is to act,[57] and one may go on to assent to a heuristics of beings in terms of potencies, forms and acts.

But, what is "explanatory knowledge by an intellectually patterned experience of the empirical residue"? What is "understanding ... in their relations to one another"? Within a generalized empirical method, we need data, we need our own instances of intellectually patterned experience of the empirical residue. There is also the challenge of being up-to-date. Otherwise, not only will one's heuristics be out of date, but one's heuristics might not be about real forms. There is the contemporary question: What is Life? As Schrödinger was aware in his time, organic life is partly physical and chemical. Recently, Addy Pross gave a more up-to-date push on the problem: *What is Life? How Chemistry becomes Biology*.[58] In order to make progress in the contemporary problem, there would seem to be no alternative but to enter the challenges posed by (and to take help provided by) ongoing developments in the modern sciences.[59]

---

[54] *CWL3*, 460-461. See also note 53, in ch. 2, above.

[55] *CWL3*, 299.

[56] Bernard Lonergan, *Phenomenology and Logic, The Boston College Lecture Series on Mathematical Logic and Existentialism*, ed. Philip McShane, *Collected Works of Bernard Lonergan*, vol. 18 (Toronto: University of Toronto Press, 2001), "Appendix A: Two Diagrams," 319-323.

[57] "It follows that potency, form and act assign not merely the structure in which being is known but also the structure immanent in the very reality of being" (*CWL3*, 525).

[58] Addy Pross, *What is Life? How Chemistry becomes Biology*. Oxford: Oxford University Press, 2012/2014.

[59] "(T)here is the rule of explanatory formulation. ... It is a rule of extreme importance,

What is Life? We say that *columbidae* is alive. And, in chapters two and three (above), a pre-heuristics was developed for multi-cellular avian life. Understanding the bird will need to be in terms of genetic sequences of multi-layerings of explanatory terms and relations, heuristically symbolized by $\int C_s(p_i; c_j; b_k; z_l)ds$. In the present context, though, we have a glimpse of a further significance of the pre-heuristics. The pre-heuristics is not only a pre-heuristics for our understanding, but a pre-heuristics of potencies, forms and acts, "components of proportionate being."[60] Each stage $C_s(p_i; c_j; b_k; z_l)$ are ;-layered aggregates of conjugate forms that are physical; chemical; and so on. It will be convenient to have a name and so might we not say that *columbidae* is *aggreformic*?

But, what are particular potencies, forms and acts of *columidae*? What are instances of "explanatory knowledge by an intellectually patterned experience of the empirical residue," and "understanding … in their relations to one another"? Can we identify physical, chemical, botanical or zoological forms of *columbidae*? And, if we speak of aggreformic (p_i; c_j; b_k; z_l) layerings, what is it about *columbidae* that shows that the bird is *alive*? What kind of being is *columbidae*?

As the sciences continue to reveal, the multi-layered *columbidae* is a complexly-multi-talented being. As helpful as it was to look to *columbidae* to make preliminary progress toward a pre-heuristics of aggreformic entities, given the complexity of multi-cellular *columbidae*, it can be helpful to look also to a simpler organism. There are, in fact, one-celled organisms that also are said to be alive. Yet, as we will find, even looking

---

for the failure to observe it results in the substitution of pseudometaphysical mythmaking for scientific inquiry. One takes the description of sensible contents, and without any effort to understand them one asks for metaphysical equivalents. One bypasses the scientific theory of color or sound, for after all it is merely a theory and, at best, probable: one insists on the evidence of red, green, and blue, or sharp or flat; and one leaps to a set of objective forms without realizing that the meaning of a form is what will be known when the informed object is understood. … Such blind leaping is inimical not only to science but also to philosophy. The scientific effort to understand is blocked by a pretense that one understands already, and indeed in the deep, metaphysical fashion. But philosophy suffers far more, for the absence of at least a virtual transposition from the descriptive to the explanatory commonly is accompanied by counterpositions on reality, knowledge, and objectivity" (*CWL3*, 528-9).

[60] See note 53.

to the one-celled organism raises challenging problems for contemporary biosciences and philosophy of biology.

4.5 The One-Celled Organism

4.5.1 *Chemotaxis*

The scientific community has known about micro-organisms for more than three centuries, thanks in part to the discoveries of Robert Hooke (1635-1703) and Antonie Philips van Leeuwenhoek (1632-1723) in the late 17th century.[61] In van Leeuwenhoek's words,

> no more pleasant sight has come before my eyes than these many thousands of living creatures seen all alive in a little drop of water moving among one another, each separate creature having its own proper motion.[62]

Certainly, van Leeuwenhoek, had no doubt that what he was seeing were "living creatures ... all alive ... moving among one another."

However, it wasn't until the 1960's that Julius Adler began a rigorous study of bacterial motion itself.[63] In 1977, Berg and Purcell published experimental and theoretical results on chemotaxis.[64] Berg and others followed up on Adler's work, and for several decades Berg has been a

---

[61] See note 44.

[62] Antonie Philips van Leeuwenhoek, "Letter 18" in, Pepper Water, *Antony van Leeuwenhoek and his "little animals"* (New York: Reprint, Dover Pubs., 1960), 144. Originally published, Clifford Dobell, F. R. S., *Antony van Leeuwenhoek and his "little animals."* New York: Harcourt Brace and Co., 1932. Also available online: https://archive.org/: https://archive.org/details/antonyvanleeuwen00dobe. Modern methods provide images of E. coli *in vivo*, at high resolution. For one example, see Anastasia I. Konokhova et al., "High-precision characterization of individual *E. coli* cell morphology by scanning flow cytometry," *Cytometry A.* vol. 83, no. 6 (June, 2013): 568-75.

[63] Julius Adler, "Chemotaxis in Bacteria," *Science* 12, vol. 153, no. 3737 (August 1966): 708-716. See also "Chemotaxis in *Escherichia coli*," *Cold Spring Harbor Symposia on Quantitative Biology*, 30 (1965): 289-292.

[64] Howard C. Berg and Edward M. Purcell, "Physics of Chemoreception," *Biophysical Journal*, Vol. 20 (1977): 193-219.

leader in this area of research.[65] There is now a large and expanding literature on the biophysics and biochemistry of the phenomenon.[66]

How is *chemotaxis* seen in the lab? A cluster of *E. coli* can be placed in a film of agar.[67] With a light microscope (1000X) and standard staining techniques, *E. coli* can be seen to move about in agar. The motion is erratic, this way and that, over short distances, and is called "run and tumble."[68] The phenomenon called *chemotaxis* results when, for instance, a small sample of sugar is placed in the agar that contains the *E. coli*.[69] The erratic style of motion continues. But, before long, something new happens. Slowly, but surely, most of the *E. coli* converge on, and consume the sample of sugar.

4.5.2 *The aggreformic one-celled organism*

Biochemistry has made great progress in identifying biochemical pathways of one-celled organisms. For instance, main metabolic pathways of *E. coli* are now well-known[70]. The complete metabolic world of *E. coli*

---

[65] http://www.rowland.harvard.edu/labs/bacteria/people_hberg.html.

[66] See, for example, Robert R. Kay, Paul Langridge, David Traynor and Oliver Hoeller, "Changing directions in the study of chemotaxis," *Nature Reviews Molecular Cell Biology*, vol. 9 (June 2008): 455-463; George H. Wadhams & Judith P. Armitage, "Making sense of it all: bacterial chemotaxis," *Nature Reviews Molecular Cell Biology*, vol. 5 (December, 2004): 1024–1037. See also: William Bialek and Sima Setayeshgar, "Physical limits to biochemical signaling," *Proceedings of the National Academy of Science, PNAS*, vol. 102, no. 29 (July 19, 2005): 10040–10045; and, H. Szurmant and G.W. Ordal, "Diversity in chemotaxis mechanisms among the bacteria and archaea," *Microbiology and Molecular Biology Reviews*, 68(2), (June, 2004):301-319; For some of the relevant experimental methods in contemporary biophysics, see, Bengt Nölting, *Methods in Modern Biophysics, Third Edition*. Berlin: Springer-Verlag, 2009.

[67] *agar*: Gelatinous material derived from algae. In biology, it is used as a medium (for example, in a petri dish) to culture bacteria and other cells in laboratory experiments.

[68] "The smooth segments … are called 'runs,' and the erratic intervals are 'tumbles'" (Howard C. Berg, *E. coli in Motion* (New York: Springer-Verlag, 2004), 33.)

[69] In the first *E. Coli* experiments by Adler, the best attractants were sugars such as galactose and glucose. Julius Adler, "Chemoreceptors in bacteria," *Science*, 166 (1966):1588-1597.

[70] See, for instance, Figure 1 in the review article, Yu Matsuoka and Kazuyuki Shimizu, "Importance of understanding the main metabolic regulation in response to the specific pathway mutation for metabolic engineering of *Escherichia coli*," *Computational and Structural Biotechnology Journal*, vol. 3, issue 4, October (2012): e201210018, 10 pages.

is, however, vast and magnificently complex. For a panoramic view of *E. coli*'s biochemistry, see one of the various indices now available in the literature.[71] In addition to alimentation pathways, there are chemoreceptor pathways[72]; pathways active in the rotation of flagella; and, indeed, altogether, there are well more than a hundred known pathways.[73] Graph-theoretic analysis reveals that "the metabolic world of this organism is not small in terms of biosynthesis and degradation"[74] (where "not small" is defined in graph-theoretic terms).

As for *columbidae*, here too, the biochemistry of *E. coli* is investigated through combinations of reaction experiments and statistical methods. After decades of research into *E. coli*, there is no evidence to suggest that we might be able to bring either cellular metabolism or spatial dynamics, of either individuals or aggregates of individuals, under some kind of systematic understanding within physics and biochemistry.[75] On the contrary, what continues to be found is that pathways of biochemical events in the cell are verifiable only in a fragmentary manner relying also on modern statistical methods. In terms of chemistry and physics,

no single model succeeds in robustly describing all of the basic elements of the cell.[76]

---

[71] *KEGG* (Kyoto Encyclopedia of Genes and Genomes), www.kegg.jp, Metabolic pathways - *Escherichia coli* K-12, MG165http://www.genome.jp/kegg-bin/show_ path way?eco01100. There is also the *Biocyc Data base collection* (2014), http://biocyc.org/.

[72] John S Parkinson, Peter Ames and Claudia A Studdert, "Collaborative signaling by bacterial chemoreceptors," *Current Opinion in Microbiology*, 8 (2005):116–121.

[73] The network includes "744 reactions that are catalyzed by 607 enzymes. ... (These) reactions are organized into 131 pathways. Of the metabolic enzymes, 100 are multifunctional, and 68 of the reactions are catalyzed by >1 enzyme. The network contains 791 chemical substrates" (from abstract of: Christos A. Ouzounis and Peter D. Karp, "Global Properties of the Metabolic Map of *Escherichia coli*," *Genome Research*, September, 10 (2000): 568-576).

[74] Masanori Arita, "The metabolic world of *Escherichia coli* is not small," *PNAS*, vol. 101, no. 6 (2004): 1543-1547.

[75] See, for example, Subrata Dev and Sakuntala Chatterjee, "Optimal search in *E. coli* chemotaxis." *Physical Review E, Statistical, Nonlinear, and Soft Matter Physics*. vol. 91, issue 4 (2015): 042714, 7 pages.

[76] M.J. Tindalla et al., "Bacterial Chemotaxis I: The Single Cell. Overview of Mathematical Approaches Used to Model, Review Article," *Bulletin of Mathematical Biology*, vol. 70

By all accounts to-date, and with high probability (judgment), chemotaxis is known to be a biased random walk, a statistical process.[77]

Still, even if *physically* and *chemically* non-systematic, perhaps chemotaxis is merely the result of statistical aggregates of merely physical and chemical acts. So, let us try to push the physics and chemistry further, if possible. When boundary conditions are sufficiently controlled, sugar samples diffuse into agar at rates that can be estimated. Also, average cellular metabolic rates of *E. coli* are known, as are various average rates in multi-scale biochemical and biophysical events.[78] Current mathematical-statistical-computational methods can be used to work out statistical cross-sections of possible distributions, including, for example, approximations to aggregate dynamics.[79] Such methods then provide diverse ranges[80] of ideal case pattern formations, in many cases approximating aggregate dynamics and attractant concentrations.[81] But,

---

(2008): 1525–1569. For source literature and experimental basis, see, for example, the references in sec. 1 of M. J. Tindalla et al.

[77] See, for example, Paul C. Bressloff, *Stochastic Processes in Cell Biology*. Cham: Springer International Publishing, Springer, 2014; and Victor Sourjik and Ned S Wingreen, "Responding to chemical gradients: bacterial chemotaxis," *Current Opinion in Cell Biology*, vol. 24, issue 2 (2012): 262–268.

[78] In addition to known biochemical pathways, cycles, average reaction-diffusion rates, and so on, there are also statistical results for biophysical events such as membrane tensions, curvatures, trajectories, ionic mass and charge distributions, and so on.

[79] This points to an extensive literature on "pattern formation" (motifs, helices, ribbons, etc.), equilibrium and non-equilibrium dynamics, space-time dependent eigen-states of such systems, and so on. See, for example, M.J. Tindalla et al., "Overview of Mathematical Approaches Used to Model, Bacterial Chemotaxis II: Bacterial Populations, Review Article." *Bulletin of Mathematical Biology*, vol. 70 (2008) 1570–1607. For particular results on, for example, traveling waves, see Jonathan Saragosti et al., "Mathematical Description of Bacterial Traveling Pulses," *Computational Biology, Public Library of Science (PLoS)*, vol. 6, issue 8 (2010).

[80] This includes: stable, non-stable, turbulent, periodic, fractal, etc.

[81] For two of many examples of recent "multi-scale" numerical results for chemotaxis, see, S. Setayeshgar et al., "Application of Coarse Integration to Bacterial Chemotaxis," *Multiscale Modeling and Simulation*, 4(1) (2005): 307–327; and Mark L. Porter, Francisco J. Valde's-Parada, and Brian D. Wood, "Multiscale modeling of chemotaxis in homogeneous porous media," *Water Resources Research*, vol. 47 (2011): W06518, 13 pages. See, also, Jonathan Saragosti, Pascal Silberzan, Axel Buguin, "Modeling E. coli

even small changes to boundary conditions can result in dramatically different experimental outcomes; and, on the mathematical side, such changes (even when small) normally require significant adjustments to models and computations. We find, then, that even for large aggregates, the statistical approach is frustrated. But, where all of these now well-known statistical results about aggregates of *E. coli* suggest that there most probably is something more than mere chemistry and physics in play, statistical results about concentrations of *aggregates* of *E. coli* do not explain the also verifiable (and non-diffusive) runs-and-tumbles of *individuals*. Evidently and self-evidently and with high probability (judgment), chemotaxis of individual cells is not fully explained – systematically or statistically – through pathways of systems of reaction-diffusion equations (whether physical, chemical, or multi-scale physical-chemical).

Advances in biophysics and biochemistry do partially explain chemotaxis. But, it is evident also that they do not tell the whole story. How, then, are we to account for, and eventually explain chemotaxis? Further clues are had by adverting to what we are getting by restricting attention to chemotaxis. In that restriction, we are getting statistical aggregates of biochemical events that, under various boundary conditions, and within allowed statistical variations, happen to fall into patterns approximated by various biochemical and biophysical pathways. But, in that restriction, are we not also abstracting from the full dynamics of a living individual *E. coli*?

It is well known that *E. coli* has many more abilities! It can scavenge for multiple food sources, to some extent it can avoid predacious protozoa and toxins. Alimentation patterns are subtle. For example, if different kinds of sugars are more or less equally available in solution, *E. coli* tends to consume available sugars in specific orders.[82] Where oxygen is

---

Tumbles by Rotational Diffusion. Implications for Chemotaxis," *One, Public Library of Science (PLoS)* (April 18, 2012).

[82] For instance: "*Escherichia coli* will often consume one sugar at a time when fed multiple sugars, in a process known as carbon catabolite repression. The classic example involves glucose and lactose, where *E. coli* will first consume glucose, and only when it has consumed all of the glucose will it begin to consume lactose. In addition to that of lactose, glucose also represses the consumption of many other sugars, including arabinose and

sometimes a chemo-attractant, it is sometimes also a chemo-repellent, depending on concentrations.[83] Some of the cell's abilities are in coordinating its movements with other *E. coli*.[84] For instance, proclivities for locomotion partly depend on the presence, or not, of other *E. coli*.[85] There are also *thermotaxis* and *phototaxis*. Associated with various anatomical parts of the one-celled organism, there are specific cellular functions such as aerobic and anaerobic respiration and excretion of metabolic products. There is predation, alimentation, gestation, and much more.

For all of these organismal functions, microbiology has been discovering complex networks of sequences, cycles and combinations of biochemical and biophysical pathways.[86] From the viewpoints of physics and chemistry, however, these pathways represent what are otherwise merely coincidental patterns of biochemical and biophysical events of described organismal functions. Nevertheless, there are patterns in the described mutual occurrences (or not) of such aggregates. Do not those patterns in data also invite explanation?

---

xylose" (Tasha A. Desai and Christopher V. Rao, "Regulation of Arabinose and Xylose Metabolism in *Escherichia coli*," *Applied and Environmental Microbiology*, vol. 76, no. 5 (March, 2010): 1524-1532.)

[83] J. Shioi, C.V. Dang and B.L. Taylor, "Oxygen as attractant and repellent in bacterial chemotaxis," *Journal of Bacteriology*, 169 (7) (July 1987): 3118-3123.

[84] In 2004, it was "demonstrated ... that under certain stress-generating conditions, cells of *E. coli* and *S. typhimurium* excrete two amino acids attractants, aspartate and glutamate. These cells then become moving sources of attractants and start interacting with each other, by coordinating chemotactic motility over a long spatial range. This interaction leads to different nontrivial collective phenomena such as formation of dense multicellular clusters, moving bands, 3D-moving structures called slugs, and complex stationary patterns" (Nikhil Mittal et al, "Motility of Escherichia coli cells in clusters formed by chemotactic aggregation," *Proceedings of the National Academy of Sciences of the United States of America*, vol. 100, no. 23 (November 11, 2003): 13259–13263.)

[85] "The tumble frequency of an individual cell strongly depends on the position of the cell within the cluster and its direction of movement. In the central region of the cluster, tumbles are strongly suppressed whereas near the edge of the cluster, the tumble frequency is restored for exiting cells, thereby preventing them from leaving the cluster, resulting in the maintenance of sharp cluster boundaries" (Nikhil Mittal et al.)

[86] See note 72.

This kind of question is not new to modern science, especially in mathematics.[87] Patterns in groupings of terms and relations can lead one to discover higher viewpoints. Here, though, we are talking about *E. coli*, so the needed higher viewpoint will not be merely mathematical. Experimental results are bringing out the need of a higher viewpoint in which we will be able to define aggregate-events in their own right. Whatever those higher terms and relations are will be found in available data. What are the available data? Data will include statistical data of otherwise merely coincidental biophysical and biochemical events; and descriptions of patterns of mutual occurrences (or not) of biological functions. Within biophysics and biochemistry, those (otherwise merely coincidental) patterns of mutual occurrences in aggregates are represented by patterned combinations of physical and chemical pathway equations. In other words, that complex symbolic data of biophysics and biochemistry also

will be clues leading to insights that pertain to the higher viewpoint of biology[88]

of the one-celled organism.

In mathematics, breakthroughs to higher viewpoints tend to be hard won, and significant. For the biology of the one-celled organism, here too we can expect a similar challenge and significance. Breakthroughs will be to (not-yet-known) correlations and terms of so-far-only-described biological functions of the one-celled organism *E. coli*.

4.5.3 *An aggreformic and more or less autonomous center of activity*

In the last section, modest progress was made toward intimating the possibility of working out a heuristics of the one-celled aggreformic organism *E. coli*. In fact, our results there also can help us move closer to

---

[87] Nor is it new to this book. See, for example, secs. 2.4-2.8, above. In the present chapter, though, the context has been expanded to included intended entities.
[88] *CWL3*, 288.

seeing that not only is *E. coli* aggreformic, but that it is a special type of central form.

Recall that, in chemical solutions, once reactants are depleted, other things being equal, reactions tend toward chemical equilibrium.[89] Something quite different happens with *E. coli*. In fact, on the cellular scale, something quite dramatic occurs. When there is no foodstuff immediately available, there soon follows a rush of biochemical activity within the one-celled organism. The cellular metabolism fires up, flagella spin, causing a whole-cell trajectory called 'a run,'[90] before its next whole-cell 'tumble' and its next whole-cell chemo-sensing in a new locale. The one-celled organism *E. coli* has both motility and the capacity to chemo-detect materials in its environment. Within ranges of boundary conditions, it is with positive (empirical) probabilities of success that cell populations are able to locate foodstuff, even when none initially are available in the immediate vicinity.

As already pointed to within our developing heuristics, *E. coli* evidently is aggreformic and, in particular, is not merely chemical and not merely physical. But, does it not seem that there is something more that is revealed in the details of how the organism encounters matter in its environment, how it locates, or avoids other living and non-living entities, how it secures its own resources?

By contrast, consider, for example, chemical compounds. Each chemical compound is defined through (all of) its capacities-to-perform within a vast chemical matrix-mesh (a basis of which is an up-to-date periodic table). Like glucose and other chemical compounds, we know that *E. coli* also has chemical properties, known through its various metabolic pathways. In a first and obvious sense, there is a lack of symmetry between the chemistry of glucose and the chemistry of *E. coli*. The chemistry of alimentation and metabolization is far more complex than glucose. For, it is understood in terms of complex combinations of pathways of physical

---

[89] Chemical equilibrium can be static or dynamic, depending partly on reversibility of reaction equations.

[90] See note 68.

and chemical equations, from cell membrane to cell interior, to excretion of chemical by-products. Still, besides such further chemical complexities, might *E. coli* otherwise be on a comparable chemical footing with glucose?

To get closer to the key issue here, let us, as it were, turn the tables on *E. coli*. Instead of thinking of *E. coli* encountering glucose, think of glucose encountering *E. coli*. There often is a reaction of sorts, which on the side of *E. coli* is called *alimentation*. This occurs through a sort of blind spontaneity of the chemically-capable cell: glucose is consumed by the cell through a large patterned aggregate of chemical events. But, alimentation often does not occur! In chemical terms, glucose is chemically reactive, or not, depending on sufficient proximity, or not, of appropriate chemical matter. But, not only does *E. coli* often not consume an available species of sugar, but *E. coli* may go on to scavenge and ultimately secure different species of sugar, or other foodstuffs eventually detected elsewhere within its environment; and go on to do many other things besides.

Even though our work here is mainly descriptive, there is the possibility of (a preliminary) detecting of something special about *E. coli*. Within a basic position,[91] this is to be done partly through our own self-detecting. Glucose, a $(p_i;c_j)$ aggreformic compound, physically; and chemically reacts, or not, as a component in a physical or chemical reaction, according to the presence or not, of suitable physical or chemical matter. There is no evidence to suggest that any species of glucose adjusts to the presence of other entities, removes itself from a possible reaction, or acts in ways that would secure the integrity of its own molecular form. On the contrary, chemical reactivities of reactants result in reactions and chemical change. Here, then, is an opportunity for self-attention. Again, for now, all of this is mainly descriptive. But, an insight is possible here, namely, that the not-alive chemical compound glucose is neither self-directed nor other-directed.

---

[91] *CWL3*, 413.

The one-celled organism *E. coli,* by contrast, is known to be not merely $(p_i;c_j)$ aggreformic. That is, the one-celled organism reveals a remarkable sensitivity and adaptability: to entities in its environment; and also to its own metabolic needs for continued survival.

As Schrödinger observed, an organism

> goes on 'doing something,' moving, exchanging material with its environment, and so forth, ... for a much longer period than we would expect of an inanimate piece of matter to 'keep it going' under similar circumstances.[92]

And, Lonergan describes the organism to be

> a more or less autonomous center of activity.[93]

What kind of being does all of this? As touched on in the previous paragraphs, in both physics and chemistry, our understanding is of laws explaining, for example, electrostatic action-and-reaction, and chemical reactants combining to yield chemical products. Our understanding in physics and chemistry is of entities that evidently are neither self-directed nor other-directed. But, there are other entities, for example the one-celled organism *E. coli,* which to various degrees are known to be able to detect the presence of other entities, to scavenge for and secure food stuff in their environment, to avoid predators, poisons, toxins and other potentially harmful entities, to eliminate otherwise damaging toxins that have penetrated cell walls, and much more. Our understanding of such an entity is of something generically different from 'neither self-directed nor other-directed.' Holding to basic position and attending to that modest insight, we can assert the existence of at least two main genera of central form. It will be convenient to have names, and so let us say *synnomic*[94] *forms,* for

---

[92] See note 4.

[93] *CWL3,* 96.

[94] Philip McShane, *The Shaping of the Foundations: Being at Home In Transcendental Method* (Washington, D. C.: University Press of America, 1976), 31, http://www.philipmcshane.ca/foundations.pdf.

non-living entities that are neither self-directed nor other-directed; and *autonomic*[95] forms for living entities.[96]

4.6 Further Questions

Section 4.5 five was a challenging foray into a few details of the life of the one-celled organism *E. coli*. In this section 4.6, a few more questions are raised, with brief discussions pointing to needed follow-up.

Relatively autonomic forms: The last section ended by taking a small step toward a heuristics of two main genera of entities - autonomic forms, living matter; and synnomic forms, non-living matter. But, besides the "many thousands of living creatures seen all alive in a little drop of water moving among one another"[97] and other known life-forms, there are also viruses, which, as noted in section 2.9· are entities "at the edge of life."[98] As revealed in X-ray crystallography,[99] outside of a host cell, viruses have a crystal-like structure (proteins and nucleic acids) and do not seem have their own metabolism. Replication cycles begin only after a virion enters

---

[95] Philip McShane, *The Shaping of the Foundations: Being at Home In Transcendental Method* (Washington, D. C.: University Press of America, 1976), 32. http://www.philipmcshane.ca/foundations.pdf. Note that the name 'autonomic' also is commonly used in a different sense, in anatomy, in reference to the *autonomic nervous system*. For example, see the discussion of the anatomy of the pigeon, in Chapter 2. An organism with an autonomic nervous system is an autonomic form. But, an autonomic nervous system is not a central form. See also McShane, *The Shaping of the Foundations*, note 92, 33.

[96] The main descriptive discussion here has barely touched the surface of the problem of intimating the possibility of beginnings in our own self-detection of our detecting detection abilities of other entities. To move beyond preliminary description of description, we will need to enter into, and work toward luminously incorporating advances made in up-to-date scientific understanding of entities, with their variously complex properties and capacities revealing that each in their own specific ways is more, or less, self-directed and/or other-directed.

[97] Antonie Philips van Leeuwenhoek, "Letter 18," note 62.

[98] E. P., Rybicki, "The classification of organisms at the edge of life, or problems with virus systematics." *South African Journal of Science*, 86 (1990):182–186.

[99] The results of X-ray crystallography are to be understood within a basic position, and a heuristics of aggreformic entities.

a host cell. To what genus of form do such entities belong? Evidently, there are entities that are neither autonomic nor synnomic. Of course, we will need to enter the details of virus science, within ongoing development in self-attention. However, does not the virus replication cycle already reveal that there is a third genus of entities, what we can call the *relatively autonomic*[100]?

Potentia activa: Through an intellectually patterned experience of the empirical residue, one will know conjugate potency, form and act.[101] But, when data are taken together, it is also possible to grasp that *E. coli* is a more or less autonomous center of conjugate activity. Are we not, then, also knowing something of central potency, an *organic potentia activa*?[102] This is not to suggest that there is evidence that Thomas Aquinas knew about nucleic acids, the metabolism of glucose, complex biomolecules or other discoveries of modern science. I am suggesting the need and possibility of an updated explanatory heuristics of *organic potentia activa*, in modern biophysics, biochemistry, botany and zoology.

Negentropy: In the present chapter, a possibility has emerged, a nudging toward a heuristics of living organisms being *autonomic* forms. But, Schrödinger's suggestions regarding the question '*What is Life?*' were along quite different lines. Searching for a way to explain the way organisms persist in apparent conflict with the Second Law of Thermodynamics, he initially suggested that a numerical *negative entropy* might be helpful, or, as it was later named by Léon Brillouin,[103]

---

[100] Philip McShane, *Shaping the Foundations*, 32-33. How many kinds of form are there? It is an empirical question. Within the broad division here called *relatively autonomic*, it is known that there are numerous species. Ongoing research is revealing new subtleties of replication cycles, and subspecies. For a recent discussion see, for example, Patrick Forterre, "Defining Life: The Virus Viewpoint," *Origins of Life and Evolution of Biospheres*, vol. 40, issue 2 (2010): 151-160.

[101] See notes 53 and 54.

[102] In a somewhat different context, Aquinas describes knowing *potentia activa*. See sec. 3.4, "*Potentia Activa*," in Bernard Lonergan, *Verbum: Word and Idea in Thomas Aquinas*, vol. 2 in *Collected Works of Bernard Lonergan* (Toronto: University of Toronto Press, 1997), 121-128.

[103] Leon Brillouin, "Negentropy Principle of Information", *Journal of Applied Physics*, vol. 24 issue 9 (1953): 1152-1163; and Léon Brillouin, *La science et la théorie de l'information*. Paris: Masson, 1959.

negentropy. However, Schrödinger himself notes that he soon was made to realize that 'entropy taken with a negative sign' does not explain, for example, why organisms need to

feed on matter 'in the extremely well ordered state of more or less complicated organic compounds' rather than on charcoal or diamond pulp.[104]

As is well known, in Boltzmann's theory, entropy S is a sum of terms of the form $S_i = - k_B p_i \log(p_i)$,[105] where each p_i is an empirical probability and k_B is Boltzmann's constant. But, statistical theories refer to ideal relative frequencies of central or conjugate acts. What about individual organisms? Even though Schrödinger's suggestions regarding a negative statistical entropy were inconclusive, do not many of his initial observations about individual organisms remain valid descriptions? Here are few examples:

Life seems to be orderly and lawful behaviour of matter, not based exclusively on its tendency to go over from order to disorder, but based partly on existing order that is kept up.

...

The general principle involved is the famous Second Law of Thermodynamics (entropy principle) and its equally famous statistical foundation.

...

What is the characteristic feature of life? When is a piece of matter said to be alive? When it goes on 'doing something,' moving, exchanging material with its environment, and so forth, and that for a much longer period than we would expect of an inanimate piece of matter to 'keep going' under similar circumstances. When a system that is not alive is isolated or placed in a uniform environment, all motion usually comes to a standstill very soon as a result of various kinds of friction; differences of electric or chemical potential are equalized, substances which tend to form a chemical compound do so, temperature becomes uniform by heat conduction. After that the whole system fades away into a dead, inert lump of matter. A permanent state is reached, in which no observable events occur. The physicist calls

---

[104] Schrödinger, "Note to Chapter 6" in *What is Life?*
[105] In the statistical formulation, small probability means large entropy.

this the state of thermodynamical equilibrium, or of 'maximum entropy.' Practically, a state of this kind is usually reached very rapidly.

...

How does the living organism avoid decay? The obvious answer is: By eating, drinking, breathing and (in the case of plants) assimilating. The technical term is metabolism.[106]

In physics and chemistry, the Second Law of Thermodynamics acknowledges a dispersiveness in physical and chemical being. In the question 'What is Life?', Schrödinger calls our attention to the abilities of organisms to, with more or less success, battle against that dispersiveness, a dispersiveness described as increasing entropy. But, where organisms overcome chemical and physical dispersiveness, when compared to physics, is not a similar supervening longevity also found in chemical entities? Verifiable in $(p_i; c_j)$ chemical elements and compounds are aggregates of physical events. But, species of chemical elements and compounds exist over much longer time intervals than can be accounted for by known dynamics of physical entities which are highly dispersive and often fleeting.[107] In other words, a chemical entity continues to exist despite a dispersiveness of merely physical being. It would seem, then, that besides its meaning in statistical theories, the word *negentropy* can name an aspect of central potency of individual entities.[108] Schrödinger's suggestions, therefore, eventually can be lifted to a verifiable heuristics where in biology is negentropic chemistry.[109]

---

[106] Erwin Schrödinger, *What is Life? The Physical Aspect of the Living Cell, with, Mind and Matter, and Autobiographical Notes.* Ch. 6, "Order, Disorder and Entropy," Ch. 6, excerpts from pars. 2, 3 and 4.

[107] Reaction rates in elementary gauge-fields are many orders of magnitude smaller than longevities of chemical elements and compounds.

[108] Philip McShane provides helpful leads on the challenge here: Philip McShane, "The Importance of Rescuing *Insight*," in, *The Importance of Insight, Essays in Honor of Michael Vertin*, eds. John J. Kiptay Jr. and David S. Liptay (Toronto: University of Toronto Press, 2007), 201; and note 13: Philip McShane, "The Conservation of Energy," *Cantower XXX* (September 1, 2004), www.philipmcshane.ca/cantower30.pdf.

[109] It seems that the relation descends in the sense that chemistry is negentropic physics; and that physics is negentropic space-time. But, I am only raising questions here, touching

Information and Communication: The word 'information' is found throughout the scientific and philosophic literature, in for example, information theories, quantum information theories, and philosophy of biology.

These days, information theories mainly focus on the mathematics of coding, sequences, symbols, computation and physical transmission.

Statistically defined entropy S and (statistically defined) negentropy (that is, $-S$) feature in contemporary theories of information and communication[110]. In contemporary philosophy of biology and computer science, communication is said to occur when there are physical or chemical events. In those contexts, the name 'information' commonly refers to what can be measured or counted.[111]

In everyday language, however, as well as in the human sciences, 'communication' is between persons. It includes gestures, expressions, signs, symbols, and mutual understandings. But, when the focus is on sense experience, it also is often said that there is communication, even when there are only symbols or signs.

In contemporary philosophy of biology, the term 'information' usually refers to secondary determinations[112] of physical and chemical acts; and when contemporary authors speak of communication occurring in physics, chemistry and biology, the emphasis would seem to be on conjugate acts. In metaphysics, information can have another meaning, such as pointed to by Lonergan:

the meaning of a form is what will be known when the in*form*ed object is understood.[113]

What is needed are detailed empirical investigations that would help us in our development toward a control of meaning about: information, communication, statistical entropy and statistical negentropy, in terms of

---

on advanced problems that will need to be handled in the context of contemporary physics; chemistry; and biology.

[110] There is, for example, the now well-known Shannon entropy. The literature is large.

[111] For some examples, see sec. 4.2, above.

[112] "Relations," sec. 16. 2 in "Metaphysics as Science," ch. 16 of *CWL3*, 514-520.

[113] *CWL3*, 528. See also note 59. Italics mine.

central and conjugate potencies, forms and acts, primary relations and secondary determinations.[114]

Mathematical modeling: Contemporary systems biology and mathematical biology are making progress partly by making use of increasingly large mathematical models that are systems of partial, ordinary and multi-scale differential equations. There are also statistical models. Typically, probability terms are added to systems of differential equations to produce what are called *stochastic* (systems of) differential equations. The added probability terms often are called "noise terms." The so-called non-stochastic systems of differential equations often are called 'deterministic.' However, in applications to actual biochemistry and biology, such differential equations usually refer to large aggregates and are verified using statistical methods. As verified, terms of such mathematical models usually represent time averages, spaces averages, or population averages, or combinations of these. Such averages are statistical terms, and in applications they obtain empirical significance when, in particular, variances also are known and statistically verified. We also find different kinds of term combined in single models. For instance, some terms may refer to types of organism (such as concentrations of *E. coli*), while other terms in the same model may refer to (densities of) conjugate events, e.g., biochemical events in the TCA cycle. While important progress is being made in systems biology and mathematical biology generally, what also is evident is a much needed control of meaning. That new control of meaning would make progress toward identifying central and conjugate potencies, forms and acts, primary relations and secondary determinations.[115]

---

[114] See note 112. To get a sense of the enormity of the challenge of including human sciences, add the word 'self' to the quotation from sec. 15.7.2, "Organic Development": "(Self-) study of an organism begins … " (*CWL3*, 489ff). The high challenge is further described in *CWL3*, sec. 17.3.8, 608-616. In particular, see "canon of explanation" (*CWL3*, 609). See also note 116.

[115] See notes 112 and 116.

4.7 "It Comes About"

The climb of the book has been revealing both the great need and the great challenge of reaching a control of meaning at the level of the times. Eventually, this will include being luminous about our knowledge of potency, form and act, components of proportiante being. In section 15.7.3 of *Insight*, Lonergan outlines "The Significance of Metaphysical Equivalence."[116] For now, the needed control of meaning evidently and self-evidently is remote to present day achievement in the sciences and philosophy of science. However, with an optimism toward a not too distant future, a good place to end this difficult chapter is with Lonergan's description, a compass bearing toward what we can expect if we continue down that road, or rather, up that climb. For, eventually, in order to be a scientist and philosopher at the level of the times, the following will be normal and normative:

> it comes about that the extroverted subject visualizing extensions and experiencing duration gives place to the subject orientated to the objective of the unrestricted desire to know and affirming being differentiated by certain conjugate potencies, forms, and acts grounding certain laws and frequencies.[117]

---

[116] *CWL3*, "The Significance of Metaphysical Equivalence," sec. 15.7.3, 530-533.
[117] *CWL3*, 537. See also frontispiece.

Chapter 5

The Concrete Intelligibility of Space and Time

Abstract: A main purpose of this chapter is to suggest beginnings toward a heuristics of the concrete intelligibility of Space and Time, referred to in the second theorem of Lonergan's Chapter 5 of *Insight*. Section 1 draws attention to the enormous complexity of the problem. Section 2 briefly recalls some of the literature from physics, on emergence. Section 3 does the same, but for biology. Taken together, sections 2 and 3 bring out further evidence of the need of foundational development. Section 4 introduces a symbolism that will be helpful in thinking about "larger and more complex questions."[1] Section 5 introduces *schemes of recurrence*, and an empirically grounded symbolism for intending history as emergent fact. Section 6 includes a discussion of probabilities of emergence of schemes of recurrence, which in turn can help us make a small beginning toward glimpsing a heuristics of *emergent probability*. Section 7 draws attention to the major problem of implementation. If generalized empirical method is desirable, it has not yet been tried. It will be difficult, partly because implementation will depend on self-attention and the emergence of a new control of meaning in science and philosophy of science. There is the question of what can we do to help promote emergence of the balanced method. There is also the question of what that future implementation will look like. These questions arise at the end of the chapter. They will be discussed briefly in the Epilogue.

---

[1] See note 72.

5.1 A Larger and More Complex Question

The reader may recall Lonergan's theorem from *Insight*, on the abstract intelligibility of Space and Time. This was already mentioned in Chapter 1, but for the convenience of the reader, is given again here:

> The abstract formulation, then, of the intelligibility immanent in Space and in Time is, generically, a set of invariants under transformations of reference frames, and specifically, the set verified by physicists in establishing the invariant formulation of their abstract principles and laws.[2]

What does the theorem mean? As discussed in Chapter 1, unpacking Lonergan's theorem will be a future challenge, at least two-fold, needing understanding in up-to-date physics and progress in control of meaning. That challenge is additionally subtle in that up-to-date meanings are part of an ongoing historical development. So, among other things, Chapter 1 of the present book also included some of Lonergan's leads on historical understanding.[3]

In addition to the theorem on the abstract intelligibility of Space and Time, there is a second theorem in Chapter 5 of *Insight*, on "The Concrete Intelligibility of Space and Time." Part of the lead up to that theorem is as follows:

> Space and Time have been defined as ordered totalities of concrete extensions and concrete durations.
>
> ...
>
> The abstract intelligibility of Space and Time ... is not so much (the intelligibility) of Space and Time, as of physical objects in their spatiotemporal relations. May one not expect an intelligibility proper to Space and proper to Time?
>
> ...
>
> What is wanted is an intelligibility grasped in the totality of concrete extensions and

---

[2] Bernard Lonergan, Insight: *A Study of Human Understanding*, eds. Frederick E. Crowe and Robert. M. Doran, vol. 2 of *Collected Works of Bernard Lonergan* (Toronto: University of Toronto Press, 1992), sec. 5.3.1, "The Theorem," 173-174. Henceforth *CWL3*.
[3] *CWL23*, 175-177.

durations and, indeed, identical for all spatiotemporal viewpoints.

...

It has been argued that a theory of emergent probability exhibits generically the intelligibility immanent in world process.[4]

Lonergan concludes the chapter with the following:

The concrete intelligibility of Space is that it grounds the possibility of those simultaneous multiplicities named situations. The concrete intelligibility of Time is that it grounds the possibility of successive realizations in accord with probabilities. In other words, concrete extensions and concrete durations are the field or matter or potency, in which emergent probability is the immanent form or intelligibility.[5]

What, though, is *emergent probability*? In section 4.2.4[6] of *Insight*, emergent probability is defined[7]; and in section 4.2.5 of *Insight*,[8] a few generic properties of emergent probability are indicated. Some of the key terms in both of those discussions are: "probabilities," "conditioned schemes of recurrence," "emergence," "survival," "stability," "development," "materials," "event," "process," "systematic" and "non-systematic."[9] Evidently, what will be needed is a developing control of meaning, not only in up-to-date physics, but also in up-to-date chemistry, biology and emergence and survival of entities. Note also that, as brought out somewhat in Chapter 4, the goal is to know not merely what *might* be, but what *is*. And so eventually the challenge will need to include transposing results to their metaphysical equivalents, that is, potencies, forms and acts.[10]

At this stage in history, the possibility of reaching an explanatory heuristics of emergent probability belongs to a somewhat distant future.

---

[4] *CWL3*, selected from 194-195.

[5] *CWL3*, 195.

[6] *CWL3*, sec. 4.2.4, "Emergent Probability," 144-148.

[7] *CWL3*, 144-151. See also Section 5.6, below.

[8] *CWL3*, sec. 4.2.5, "Consequences of Emergent Probability," 148-151.

[9] Some of these will be discussed in some detail, in sections 5.5 – 5.8, below.

[10] "Such transposition may be easy or difficult, but insofar as it is found difficult, there also will be found some measure of ignorance taking cover under the abstract expression" (*CWL3*, 527).

The work of the previous four chapters helps bring out some of the difficulties and possibilities. The totality includes all births, lives and deaths of all genera and species, past, present and emergent; all of the airways, oceans, streams, ponds, forests and jungles of the world, fresh and polluted; the Concierto de Aranjuez for guitar and orchestra, composed in Braille by the almost blind Joaquín Rodrigo,[11] in fact, all great art and all low education; military groups and the world drug trade; all instances of cultural development as well as cultural decline.[12]

Astrophysics and astrochemistry tell us that the universe probably began with some kind of a Big Bang[13] approximately 13.8 billion years ago; followed by subsequent formations of galactic gases and dust; later, stars; and the now approximately 100 - 200 billion galaxies, each of which consists of stars, comets, asteroids, planets and other formations. The story of our own planet reaches back approximately 2.5 billion years; to early chemical compounds; to the emergence of flowers that changed the world[14]; on through to various series of plant and animal kingdoms, their historical biogeography,[15] and their extinctions; it includes the emergence of early human groups; indeed, all of human history. In present times, it includes a now global humanity possessing what in many ways are impressive skills, but also struggling with profound difficulties.

What, then, is the "immanent form or intelligibility"[16]? As pointed to above, we can take some advantage of (dense) doctrinal pointers that are

---

[11] Joaquín Rodrigo (1901-1999).

[12] For helpful detail, see *CWL3*, sec. 7.8.

[13] See, for example, Simon Singh, Simon. *Big Bang: The Origin of the Universe*. New York: Fourth Estate, 2004.

[14] For a somewhat dated, but in its essential features still relevant, description, sèe, Loren Eiseley, "How Flowers Changed the World," ch. 5 in, Loren Eiseley, *The Star Thrower*, *Introduction by W. H. Auden* (New York: Harcourt, 1978), 66-75. However, "(n)ew fossils push the origin of flowering plants back by 100 million years to the early Triassic." *Science Daily*, October 2013, http://www.sciencedaily.com/releases/2013/10/131001191811.htm. For the original article, see Peter A. Hochuli and Susanne Feist-Burkhardt, "Angiosperm-like pollen and Afropollis from the Middle Triassic (Anisian) of the Germanic Basin (Northern Switzerland)," *Frontiers in Plant Science*, vol. 4 (2013).

[15] Douglas J. Futuyma, *Evolution*, 2nd ed., (Sunderland, MA: Sinauer, 2009), ch. 6, "The Geography of Evolution," 133-159.

[16] *CWL3*, 195.

sections 4.2.4 and 4.2.5 of *Insight*; as well as Lonergan's other writings on emergent probability (in *Insight* and throughout the *opera omnia*). Explanatory heuristics at the level of the times are not presently within reach. However, something can be done now toward a pre-heuristics. And, present descriptive searching can give us hints toward our future growth.

5.2 Contemporary Physical Sciences on Emergence

The totality, of course, includes what is being discovered by contemporary physics. Although, even with that focus, fundamental difficulties soon arise. According to contemporary physics, there have been fourteen main cosmological epoch since the Big Bang: A first is the Planck Epoch, zero to approximately 10^{-43} seconds[17]; a second is called the Grand Unification Epoch, from 10^{-43} seconds to 10^{-36} seconds[18]; ...; the eighth is the Nucleosynthesis Epoch,[19] from 3 minutes to 20 minutes. There are six more epochs, with times scales increasing from millions of years, then to billions of years. The fourteenth and present epoch is the Solar System Formation Epoch, 8.5 - 9 billion years after the Big Bang, in other words, the most recent 4.5 to 5 billion years. The Sun of our Solar System sometimes is called "a late-generation star," for it incorporates matter from many earlier generations of stars. In fact, the entire universe continues to cycle and recycle available matter.

We have a sequence, not stages of organismal development, but epochs along a cosmological timeline. As just mentioned, the physics community has been making progress in sorting out dynamics and structurings of the physics and physical chemistry of the whole sequence - transitions, propensities and probabilities. But, contemporary physics also is puzzling over the contemporary Standard Model, with a long list of unsolved

---

[17] The universe spans a region of approximately 10^{-35} meters. The temperature is over 10^{32} Kelvin.

[18] There are gravity, fundamental forces and the earliest elementary particles.

[19] The temperature of the universe falls to about a billion degrees; protons and neutrons emerge; fusion and emergence of simple elements like hydrogen, helium and lithium. After about 20 minutes, the temperature and density of the universe has fallen to where the initial nuclear fusion stops.

fundamental problems. Furthermore, there are no signs yet of an emerging consensus regarding the nature of emergence, in particular, emergence associated with the beginnings and evolution of life on earth.[20]

In *Science and Ultimate Reality, Quantum Theory, Cosmology, and Complexity,*[21] Clayton's expository article includes the following:

> The emergence hypothesis requires that we proceed through at least four stages. The first stage involves rather straightforward physics – say, the emergence of classical phenomena from the quantum world (Zurek 1991, 2002) or the emergence of chemical properties through molecular structure (Earley, 1981). In a second stage, we move from the obvious cases of emergence in evolutionary history toward what may be the biology of the future: a new law-based "general biology" (Kauffman 2000) that will uncover the laws of emergence underlying natural history. Stage three of the research program involves the study of "products of the brain" (perception, cognition, awareness), which the program attempts to understand not as unfathomable mysteries but as emergent phenomena that arise as natural products of the complex interconnections of brain and central nervous system. Some add a fourth stage to the program, one that is more metaphysical in nature: the suggestion that the ultimate results, or the original causes, of natural emergence transcend or lie beyond Nature as a whole.[22]

Clayton also draws attention to a list of

> empirical differences that are reflected in these diverse senses of emergence: temporal or spatial; progression from simple to complex; increasingly complex levels of information processing; new properties (physical, biological, psychological); causal entities (atoms, molecules, cells, central nervous system): new organizing principles or degrees of inner organization (feedback loops, autocatalysis, autopoiesis); and the development of "subjectivity."[23]

---

[20] M. Gargaud et al., eds., *From Suns to Life: A Chronological Approach to the History of Life on Earth,* Springer ebook, 2006. (Reprinted from the journal, *Earth, Moon and Planets,* vol. 98, issue 1-4, June, 2006.)

[21] John D. Barrow and Paul C. Davies and Charles L. Harper, Jr., eds., *Science and Ultimate Reality, Quantum Theory, Cosmology, and Complexity.* Cambridge: Cambridge University Press, 2004.

[22] Philip D. Clayton, "Emergence: us from it," ch. 26 in, Barrow, Davies and Harper, 578. Parenthetic references are in source document.

[23] Clayton, 579.

Some examples are drawn from "artificial systems, biochemistry, biology and neuroscience."[24] Others are from fluid dynamics, simulated evolutionary systems (computer games called "cellular automata"), ant colony behavior, the biochemistry of cell aggregation and differentiation, and local-global interactions.

 Concluding his article, Clayton comments:

> To the extent that the evolution of organisms and ecosystems evidences a "combinatorial explosion" (Morowitz 2002), ..., the hope of explaining entire living systems in terms of simple laws appears quixotic.[25]

..

> Ultimately, emergence involves the prediction that increases in complexity will correlate with specific transition points where new types of structural organization will appear." [26]

 In the article following Clayton's expository work, Ellis goes on to speak of

> true complexity, ..., to be distinguished from what is covered by statistical physics, catastrophe theory, study of sand piles, the reaction-diffusion equation, cellular automata, ..., and chaos theory.

...

> True complexity involves vast quantities of stored information and hierarchically organized structures that process information in a meaningful manner, particularly through implementation of goal-seeking feedback loops. Through this structure they appear purposeful in their behavior ("teleonomic"). This is what we must look at when we start to extend physical thought to the boundaries, and particularly when we try to draw philosophical conclusions – for example, as regards the nature of existence – from our understanding of the way physics underlies reality. Given this complex structuring, one can ask, "What is real?", that is, "What actually exists?", and "What kinds of causality can occur in these structures?

...

---

[24] Clayton, 580.
[25] Clayton, 597.
[26] Clayton, 604.

Complex emergence is enabled ... according to the system structure.

...

The essential point of systems theory is that the value added by the system must come from the relationships between the parts, not from the parts per se (Emery 1972; von Bertalanffy 1973). ... The principles of hierarchy and modularity have been investigated usefully in the context of computing, and particularly in the discussion of *object-oriented programming* (see Booch 1994), and it is helpful to see how these principles are embodied in the physical and biological structures. [27]

Near the end his paper, Ellis provides a column representing a "hierarchy of causal relations." Metaphysics is at the base, immediately above which is "Theory of Everything," then Particle Physics, and then Chemistry. Continuing upwards beyond Chemistry, the chart splits into two columns: Materials on the left with Biochemistry on the right; above that are Geology on the left and Physiology on the right; then Astronomy on the left and Psychology on the right; Cosmology on the left and Sociology on the right; and in the top row, Metaphysics (again) on the left and Ethics on the right.[28] The

causal hierarchy rests on a metaphysical ultimate reality as indicated ... Here the unknown metaphysical issues that underlie both the choice of specific laws of physics on the one hand, and specific conditions for cosmology on the other, are explicitly recognized.[29]

Problems are evident, and various questions arise. Is there anything "straightforward"[30] about the emergence of classical phenomena from the quantum world?[31] What are *molecular structure* and *chemical properties*[32]? Is it coherent with experience to suggest that "products of

---

[27] George F.R. Ellis, "True Complexity and its associated ontology," ch. 27 in, *Science and Ultimate Reality, Quantum Theory, Cosmology, and Complexity*, eds. John D. Barrow and Paul C. Davies and Charles L. Harper, Jr. (Cambridge: Cambridge University Press, 2004), selections from 607-610.

[28] Ellis, Figure 27.8, 634.

[29] Ellis, 634.

[30] See note 22.

[31] See note 22.

[32] See note 22. The reference in Ellis' paper is to Earley's work: "The diiodine molecule, I_2,

the brain (perception, cognition, awareness)"[33] are merely emergent phenomena[34]? Is there, perhaps, performance-contradiction? Is a central nervous system a causal entity? Do organisms "process information"? And so on.

What we find, in fact, are various clashing views and methods. Clayton invites attention to empirical differences. By contrast, Ellis sets up a metaphysical chart: "What is 'ultimate reality'?" Ellis "emphasized ...

---

is composed of two iodine atoms joined by a covalent bond. Each of the two iodine atoms is composed of a positively-charged nucleus, a number of core electrons which comprise filled electronic shells, and an unfilled (valence) shell of electrons. The covalent bond arises from delocalization of the valence-shell electrons of both atoms, so that all these electrons may be regarded as pertaining to the molecule as a whole. The result of this bond-formation is that there is a single value of internuclear distance which is a defining characteristic of the molecule of diiodine. If, for any reason, the distance between the two iodine-atom nuclei should be less than that distance, there will be an unbalanced force (mainly arising from interactions of the filled shells) which will tend to increase the distance between the atomic centers. Conversely, if the atoms should be further apart than this equilibrium internuclear distance, then there would be an unbalanced force (primarily due to valence-shell interactions) which would tend to decrease the interatomic distance. At ordinary temperatures diiodine molecules vibrate continuously around the equilibrium internuclear distance, alteratomic distance" (Joseph E. Earley, "Self-Organization and Agency In Chemistry and In Process Philosophy," *Process Studies, Special Issue on Whitehead and Natural Science*, vol. 11, issue 4 (Winter 1981): 242-258. See Section II, "Structure and Stability in Chemistry"). Evidently, there are problems here calling out for control of meaning and progress toward determining metaphysical equivalence. See Section 4.7. Among other things, as discussed by Earley, the diiodine molecule is thought to be (a non-verifiable) imaginable aggregate of valence shells, electrons and nuclei. In the last paragraph of Earley's Section II, this view is extended to all molecules: "Just as the diiodine molecule vibrates about its equilibrium internuclear distance, and molecules in solid iodine jiggle around in three dimensions, still keeping the overall structure of the crystal intact, so, too, there are fluctuations in the concentrations of all chemicals which participate in chemical equilibria. In each small region of space the concentration of a given reagent is not quite constant, but undergoes increases and decreases, usually small." The control of meaning needed will include doubly aggreformic chemical entities of the generic form $(p_i; c_j)$. This is not to single out Earley's work. The needed control of meaning will be a major development in science and philosophy within an implemented generalized empirical method.

[33] See note 22.

[34] See note 22.

complex systems, where information plays a central role in their emergence."[35] But, as already discussed in Chapter 2, contemporary philosophical views regarding information and systems tend to be non-verifiable conceptual, logical and mathematical constructs. At the same time, although results are mixed,[36] systems theorists are making notable progress working out complex combinations of dependencies and conditional probabilities of events and occurrences - in the early universe; ecosystems, evolution; and in biological development.[37]

As Clayton points out,

(i)n the broader discussion the term 'emergence' is used in multiple and incompatible senses, some of which are incompatible with the scientific project.[38]

Kauffman also comments on the general lack of consensus regarding emergence:

Note that we have, as yet, no developed language in physics, chemistry, or biology, to discuss these matters. Consider also the miracle of the cell building a copy of itself, then the two repeat the process to make four cells, then eight, then a bacterial colony. I can only stumble with ordinary English: the cell achieves a propagating organization of building, work, and constraint construction that completes itself by the formation of a second cell. Is this matter alone, energy alone, entropy alone, or information alone? No. Do we have a formulated concept for what I have just described? No.

Yet just such propagating organization occurs. ... We have no mathematical framework that I know of, which captures this process. It appears to be a new state of matter – call it a living state.[39]

---

[35] Ellis, 634-635.

[36] *CWL3*, 514.

[37] Ellis, 615-616.

[38] Clayton, 601.

[39] Stuart Kauffman, "Autonomous Agents," ch. 29 in Barrow Davies and Harper, 662. The present literature on emergence is extensive, a complex mesh of views, from the more philosophic to the more scientific. See, for example: References in, Michael Silberstein and Anthony Chemero, "After the philosophy of mind: Replacing Scholasticism with science," *Philosophy of Science* 75 (2009): 1–27; Philip Clayton, *Mind and Emergence: From Quantum to Consciousness.* New York: Oxford University Press, 2004; Terrance Deacon, "Emergence," ch. 5 in *Incomplete Nature: How Mind Emerged from Matter.* New

5.3 Contemporary Life Sciences on Emergence

The black-throated honeyguide (*Indicator indicator*)

> cruises the forest hunting a beehive, then, chattering and darting, leads the badger to its find. The clawed beast tears apart the tough walls of the hive, and both the honey guide and badger share in a wax-comb-and-honey repast.[40]

Elsewhere, orchid flowers mimic insect pheromones, that way attracting bees, wasps, hoverflies and other pollinators.[41]

York: W. W. Norton & Co., 2012; and also Timothy O'Connor and Hong Yu Wong, "Emergent Properties," *The Stanford Encyclopedia of Philosophy* (Summer 2015 Edition), Edward N. Zalta (ed.), first published Sep 24, 2002; substantive revision June 3, 2015, http://plato.stanford.edu/archives/sum2015/entries/properties-emergent/. For one example, I draw attention to the work of Paul Humphreys, on 'emergence as fusion,' in which events are unified multi-level events. His symbolism for hierarchies of levels $L_i < L_j$ somewhat resembles symbolism provided in the present book. However, Humphrey's arguments are given in a context of philosophical discussion that becomes entangled in logical difficulties. For the time being, "(s)uch worries link back to the issue of the extent of the applicability of the view" (O'Connor and Wong, sec. 3.2.2). I note, in particular, that the logical structures $L_i < L_j$ are remote to individual living organisms, and do not draw on or connect with detailed empirical results in scientific practice, such as biochemistry, biology and evolution. Within a generalized empirical method, there will be progress toward a precise empirically informed heuristics of aggreformism. Holding to a basic position, and developing within the scientific context, higher biological conjugate forms will be known to be "emergent in underlying neural configurations or dispositions as insights are emergent in images and functions in organs" (*CWL3*, 493). And so the book *Incomplete Nature: How Mind Emerged from Matter* (Deacon, *Incomplete Nature*), that consists mainly of discussion about neural complexity and organization, falls short of the empirical challenge. In particular, it ignores a primary source of data on the problem, namely, data of consciousness. This is not doing justice to the extensive scholarship of Deacon. It is telling of the present ethos, however, that none of *wonder, insight, understanding, judgement, deliberation* or *decision* are indexed or discussed in a book on *mind*.

[40] Howard Bloom, *Global Brain, The Evolution of Mass Mind from the Big Bang to the 21st Century* (New York: John Wiley & Sons, 2000), 207.

[41] See, for example, Johannes Stökl et al., "Smells like aphids: orchid flowers mimic aphid alarm pheromones to attract hoverflies for pollination," *Proceedings of the Royal Society B*, 278 (2011): 1216–1222.

During the fall (in the northern hemisphere), eastern North American monarch butterflies (*Danaus plexippus*) migrate south from their northern range, to overwinter in sites atop the mountains of Michoacán in central Mexico.[42]

Many butterfly populations travel more than 4000 km. It has been known for some time that they manage this incredible journey partly through a sensitivity to changes in circadian sunlight patterns that match latitude and direction.[43] It is now also known that monarchs have "an inclination magnetic"[44] sensitivity that is another factor in sustained directionality, especially important during times of diminished sunlight. In fact, contemporary biology, ecology and environmental science have been learning about multitudes of symbiotic relationships and mutual dependencies, in above ground environments, underwater environments, and subterranean environments. These include local and global biochemical and predator-prey patterns, intercontinental migratory patterns; global carbon, nitrogen and water cycles; global meteorological patterns, and, indeed, the complex mutual local and global functionings of the entire biosphere.

---

[42] S. M. Perez, O. R. Taylor and R. Jander, "A sun compass in monarch butterflies," *Nature* 397 (1997): 29.

[43] Perez et al., 29.

[44] Patrick A. Guerra, Robert J. Gegear and Steven M. Reppert, "A magnetic compass aids monarch butterfly migration," *Nature Communications* 5 (June, 2014): article no. 4164. See also, Stanley Heinze and Steven M. Reppert, "Sun Compass Integration of Skylight Cues in Migratory Monarch Butterflies," *Neuron* 69 (Jan. 2011): 345–358. This has up-to-date results (2011) and bibliography. Although, the reference is given with a caveat: The empirical results in the article are clouded within a systems biology viewpoint. This is illustrated, for example, by the following: "In general, this sun compass mechanism postulates that skylight cues, providing directional information, are sensed by the eyes and that this sensory information is then transmitted to a sun compass system in the central brain. There, information from both eyes is integrated and time compensated by the circadian clock so that flight direction is constantly adjusted to maintain a southerly bearing over the day" (Stanley Heinze and Steven M. Reppert, 345). There is the familiar problem of asserting the existence of non-verifiable "information" and "systems." This problem is not to be resolved easily. Here we have another illustration of positive results that will need to be recycled within a future control of meaning. Efficient and effective recycling will occur within a future functional collaboration. See Epilogue.

However, it is also known that over the millennia the biosphere has been changing. The "History of Life on Earth" [45] includes main epochs called Precambrian, Paleozoic, Mezozoic and Cenozoic, within which there have been eras of "diversification and extinction of major groups of organisms,"[46] along with changes in global climate and geology.

Darwin's[47] results led to a shift in heuristics of the natural world. In his initial field reports, he wrote of "transmutation,"[48] "evolution"[49] and "natural selection."[50] But, in the 20th century there were major advances in biochemistry, including the mid-century discovery of heritable deoxyribonucleic acid (DNA), by Crick and Watson.[51] Ongoing advances in biochemistry, paleontology and the emerging science of ecology all have been contributing to the hard won, and now generally accepted, theory of evolution called the Modern Evolutionary Synthesis.[52]

The Modern Evolutionary Synthesis is summarized differently by different authors. However, it is basically stable, with commonalities, one of which is that it is *gene-centered*. According to the Modern Synthesis, evolution is explained in terms of theoretical "gene flow" (between different species, e.g., cross-pollination), "genetic drift" (in a species, e.g. alleles and skin pigmentation) and "genetic mutation" (in a species, e.g., due to exposure to radiation, or toxic chemicals).

In her discussion of the modern synthesis, West-Eberhard observes:

> In general, ideas that are obviously compatible with the fundamentally genetic theory of evolution, such as the theory of speciation by geographic, gene-pool

---

[45] Douglas Futuyma, ch. 5.

[46] Futuyma, 101.

[47] Charles Darwin (1809-1882).

[48] See, for example, Bryson Brown, *Evolution, A Historical Perspective* (Westport, CT: Greenwood Press, 2007), chs. 3 and 4.

[49] Brown, chs. 3 and 4.

[50] Brown, chs. 3 and 4.

[51] J. D. Watson J. D and F. H. Crick, "Molecular structure of nucleic acids. A structure for deoxyribose nucleic acid." *Nature* 171 (4356) (April, 1953): 737–738.

[52] See, for example, Futuyma, 9-11.

isolation and the roles of mutation and genetic drift in the initiation of novelty, have been emphasized. [53]

However, West-Eberhard and other scholars have been calling the Modern Evolutionary Synthesis into question. For example, Shapiro writes the following:

> Like the man searching for his key under the lamppost, we currently focus our thinking about heredity almost completely on DNA sequences, because our ability to read and manipulate them lies at the heart of present-day biotechnology. Nonetheless, we should never forget that not all heredity involves the transmission of nucleotide sequences in DNA and RNA molecules. To date, all studies of genetically modified organisms have required an intact cell structure for the introduction of new genetic information by DNA or nuclear transplantation. ... As Rudolph Virchow articulated it in 1858, "*omnis cellula e cellula*" ("every cell comes from a cell").[54]

West-Eberhard goes further:

> The result so far has been a piecemeal synthesis and a heightened awareness that the modern neo-Darwinian approach may be incomplete, without a broad treatment from the point of view of evolutionary biology as a unified science.
>
> ...
>
> Darwinians after Darwin have embraced a whole set of interrelated ideas, which now forms a coherent package of mutually reinforcing concepts. Their internal consistency imparts a kind of selective blindness to facts that do not fit.[55]

West-Eberhard goes on to discuss various inconsistencies of the modern synthesis.

> Perhaps the most remarkable inconsistency, and the most urgent to resolve, is the practice of phrasing evolutionary explanations almost entirely in terms of genes,

---

[53] Mary Jane West-Eberhard, *Developmental Plasticity and Evolution* (Oxford: Oxford University Press, 2003), 6.
[54] James A. Shapiro, *Evolution, A View from the 21ˢᵗ Century* (Upper Saddle River, NJ: FT Press, 2011), 3.
[55] West-Eberhard, 5-6.

while the phenomena we endeavor to explain are phenotypes, always products of both environment and genes.[56]

Shapiro and West-Eberhard have fundamentally different viewpoints. West-Eberhard looks to development, evolution and phenotypes, while Shapiro emphasizes computational systems. But, it is not the purpose here to enter into a detailed comparison of their viewpoints. It is enough to note their shared concern that a gene-centered theory of evolution cannot be the whole story.

A similar concern is expressed by Tooby and Cosmides:

> It would be a coincidence of miraculous degree if a series of … function-blind events, brought about by drift, by-products, hitchhiking, and so on, just happened to throw together a structure as complexly interdependently functional as the eye.[57]

One of the contemporary approaches intended to help resolve these issues is to investigate not just individual organisms but ecosystems (of living and non-living matter, also called biotic and abiotic factors). But, what is an *ecosystem*? Applications of the word include pond, lake, forest, as well as complexly related collections of sectors of the biosphere.[58] A standard contemporary definition of *ecosystem* is:

> A biological community and the physical environment associated with it. Nutrients pass between the different organisms in an ecosystem in definite pathways; for example, nutrients in the soil are taken up by plants, which are then eaten by herbivores, which in turn may be eaten by carnivores (Food Chain). Organisms are classified on the basis of their position in an ecosystem into various trophic levels. Nutrients and energy move round an ecosystem in loops or cycles (in the case above, for example, nutrients are returned to the soil via animal wastes and decomposition). See also Carbon Cycle; Nitrogen Cycle.[59]

A less detailed definition is:

---

[56] West-Eberhard, "The Genotype-Phenotype Problem," 16.

[57] J. Tooby and L. Cosmides, "The psychological foundations of culture," ch. 1 in, *The Adapted Mind* (Oxford: Oxford University Press, 1992), 57.

[58] Frank B. Golley, *A History of the Ecosystem Concept in Ecology, More than the Sum of the Parts*. New Haven, CT: Yale University Press, 2012.

[59] Elizabeth Marin and Robert Hine, eds., *A Dictionary of Biology*, 6th ed. Online Version: Oxford University Press, 2014.

An *ecosystem* is the interacting system made up of all the living and non-living objects in a specified volume of space.[60]

Looking to an ecosystem in which an organism survives certainly tells a more complete story than merely knowing about DNA. However, by definition, an ecosystem is a community within the biosphere (a pond; a lake; a valley; a forest; a desert; a continental region; a multi-continental region; and so on). And, it is known that biotic and abiotic mutual dependencies go beyond any single community. So it is that, when focusing attention on any one ecosystem, either implicit or explicit, there is always an hypothesis of the form 'other things being equal.'[61]

In an attempt to accommodate all cases, Systems Ecology extrapolates basic premises of systems biology, from organisms to ecosystems. In systems ecology, mathematical models and computer simulations are developed for selected aspects of an ecosystem. The viewpoint is that, in principle, sufficiently inclusive models of that type will explain the entire biosphere.[62] The systems approach has the advantage of going further than merely working out gene sequences, and has been making progress in understanding dynamic mutual dependencies in ecosystems.

---

[60] Kathleen C. Weathers, David L. Strayer, Gene E. Likens, *Fundamentals of Ecosystem Science* (Salt Lake City, UT: Academic Press, 2012), 3.

[61] In evolution, among other things, one investigates changes in organisms present in ecosystems, changes in 'other things equal.'

[62] Kevin J. Purdy et al., "Systems Biology for Ecology: From Molecules to Ecosystems," ch. 3 in *Integrative Ecology: From Molecules to Ecosystems*, vol. 43 in *Advances in Ecological Research* (2010): 87-149. "A systems approach to ecology ultimately requires the ability to scale from molecules to ecosystems, but making these connections is an extremely challenging task" (Kevin J. Purdy et al., Part VII. Linking Across Multiple Levels of Organisation: The Key to Understanding the System, Section VII A. Scaling from Molecules to Ecosystems, 122). "Systems ecology: Branch of ecosystem ecology (the study of energy budgets, biogeochemical cycles, and feeding and behavioral aspects of ecological communities) that attempts to clarify the structure and function of ecosystems by means of applied mathematics, mathematical models, and computer programs. It concentrates on input and output analysis and has stimulated the development of applied ecology: the application of ecological principles to the management of natural resources, agricultural production, and problems of environmental pollution" (Encyclopedia Britannica, systems-ecology).

For some, the systems approach has also been part of a Gaia Hypothesis,[63] whereby the earth is thought to be a self-regulating macro-system. There also has been a derived hypothesis that the biosphere is a Global Brain.[64] In that view, the Global Brain is thought to be a large machine.[65] Details of Lovelock's initial version of the Gaia Hypothesis have not been supported by empirical results. However, one of the underlying ideas, that of minding the whole earth, is of growing interest to the scientific community. And so modified versions of the Gaia Hypothesis are now being investigated within mainstream science.[66] Regarding the Global Brain hypothesis, as discussed in previous chapters, the notion of "computable system," "information systems," and the like are non-verifiable even in one-celled organisms, let alone evolution of global ecosystems. At the same time, it is partly thanks to multi-scale mathematical population dynamics that contemporary environmental science is making considerable progress in understanding statistically verifiable cycles, feed-back loops, and other mutual dependencies in world ecosystems. The study of such interdependencies has become a main focus in contemporary ecology.

---

[63] J. E. Lovelock, "Letter to the Editors, Gaia as Seen through the Atmosphere," in vol. 6 of *Atmospheric Environment* (Oxford: Pergamon Press, 1972) 579-580.

[64] Howard Bloom, *Global Brain, The Evolution of Mass Mind from the Big Bang to the 21st Century*. New York: John Wiley and Sons, 2000.

[65] Some of the problems here already have been discussed in Chapter 2, above. "Social animals are linked in networks of information exchange. Meanwhile, self-destruct mechanisms turn a creature on and off depending on his or her ability to get a handle on the tricks and traps of circumstance. The result is a complex adaptive system – a web of semi-independent operatives linked to form a learning machine. ... Our pleasures and our miseries wire us humans as modules, nodes, components, agents and microprocessors in the most intriguing calculator ever to take shape on this earth. It's the form of social computer which gave not only us but all the living world around us its first birth" (Howard Bloom, 12-13).

[66] See, for example, Axel Kleidon, Yadvinder Malhi and Peter M. Cox, "Maximum entropy production in environmental and ecological systems," *Philosophical Transactions of the Royal Society B*, 365 (2010): 1297–1302. In this paper, the study is on "the Gaia hypothesis, ..., the biosphere and the wider Earth system"; "complex systems ranging from the living cell to planet Earth." (Axel Kleidon et al.), 1297.

Throughout, there are basic problems that already have been discussed in Chapter 2. In the present context, note, for example, that to determine genetic sequencing, one appeals to experimental results on cell samples, with results remote to a living organism.[67] This is not to suggest that the biochemistry of genetic sequences does not contribute to an explanatory understanding of organisms. The discovery of DNA by James Watson and Francis Crick was a major breakthrough[68] in biology. The DNA of an organism limits organic possibilities to ranges compatible with biochemical capacities of a particular DNA structure, and provides a biochemical basis for continuity of species. There is also a class of homeobox genes that reveals continuities across genera.[69] But, as already brought out in chapters 2, 3 and 4, it is evident also that even one-celled organisms are not merely chemical. Whether one is working with a gene-centered theory of evolution, or mathematical models for multi-species ecosystems, the empirical evidence is that biochemical compounds and organisms are aggreformic entities, and in particular, are not aggregates of, for example, chemical entities.

What we find, then, are variously clashing points of view. Taken together, these provide further evidence of the need for a more adequate control of meaning. And, as discussed in Section 4.4, development in control of meaning will include progress toward determining metaphysical equivalents[70] of (experimental results on what are called) 'genetic drift'

---

[67] Elizabeth A. Zimmer and Eric H. Roalson, eds., *Molecular Evolution: Producing the Biochemical Data*, vol. 395, Part B, in *Methods in Enzymology*. Amsterdam: Elsevier Academic Press, 2005. Genetic sequencing is obtained through complex combinations of chemical and physical lab techniques that involve isolated proteins, amino acids, codons and nucleotides.

[68] The famous original article is: J.D. Watson and Crick, F. H. Crick, "A structure for deoxyribose nucleic acid," *Nature*, vol. 171 (1953): 737–738. A brief historical account is: Leslie A. Pray, "Discovery of DNA structure and function: Watson and Crick," *Nature Education* vol. 1, issue 1 (2008): 100.

[69] Walter J. Gehring, "The homeobox in perspective," *Trends in Biochemical Sciences*, vol. 17, issue 8 (1992): 277 - 280. The paper is on the fact that homeobox class of genes "seems to be much more universal than originally anticipated" (Gehring, 277). But, the paper also speaks of "master control genes" (Gehring, 277), which is non-verifiable. In the present book, the fiction of such control already has been discussed.

[70] See notes 10 and 32.

and 'genetic flow.' Heuristics will be of aggreformic[71] organisms living in environments that include the earth, the sun and beyond.

5.4 Symbolism

The last section ended with comments on the emerging need of a reaching and verifiable heuristics. Lonergan's advice given to some of his students will be helpful here:

> The aim of discursive reasoning is to understand; and it arrives at understanding not only by grasping how each conclusion follows from premises, but also by comprehending in a unified whole all the conclusions intelligibly contained in those very principles. Now this comprehension of everything in a unified whole can be either formal or virtual. It is virtual when one is habitually able to answer readily and without difficulty, or at least 'without tears,' a whole series of questions, right up to the last 'why?' Formal comprehension, however, cannot take place without a construct of some sort. In this life we are able to understand something only by turning to phantasm; but in larger and more complex questions, it is impossible to have a suitable phantasm unless the imagination is aided by some sort of diagram.[72]

In the previous chapters, a preliminary symbolism for a pre-heuristics of aggreformic *columbidae* was developed: $C_s(p_i; c_j; b_k; z_l)$. This was obtained by appealing to biology. Along the way, other cases were indicated. For example, a verifiable heuristics for chemical entities will be of the form $(p_i; c_j)$.[73] In Chapter 4, considerable effort was devoted to the

---

[71] Recall that an aggreformic entity, a central form, has layerings of conjugate forms. See secs 4.4 – 4.7.

[72] Bernard Lonergan, *The Ontological and Psychological Constitution of Christ*, vol. 7 in, *Collected Works of Bernard Lonergan* (Toronto: University of Toronto Press, 2002), 151. This is an empirical claim. It is for the reader to verify in one's own experience. The context of Lonergan's advice includes modern science and philosophy of science. Centuries earlier, in less developed contexts, both Aristotle and Thomas Aquinas made similar empirical claims: "The faculty of thinking then thinks the forms in the images" (Aristotle, *De Anima*, trans. J. A. Smith, The *Internet Classics Archive, MIT*, Book III, Part 7, par. 5, http://classics.mit.edu/Aristotle/soul.3.iii.html); and, "I answer that, in the present state of life in which the soul is united to a passible body, it is impossible for our intellect to understand anything actually, except by turning to the phantasms" (Thomas Aquinas, *Summa Theologica*, Part I, Question 84, Article 7).

[73] See note 32.

one-celled organism *E. coli*. By appealing to contemporary biology, we find that the one-celled organism is a living entity with talents not fully explained by physics and chemistry. Within the context of this introductory book, however, neither for *E. coli* nor for *columbidae* was there an attempt to work out what the higher biological correlations might be that would define specific biological talents for either species. Those are empirical questions and will be for the biology community to discover within a developing control of meaning.

For this fifth chapter, however, there is a "larger and more complex"[74] question, namely, the intelligibility of the totality.

The totality that includes us, the human organism. The human organism? The searchings of the first four chapters invited beginnings in self-climbing through a few skimpily treated details in biology. As may be increasingly evident, it will be a major achievement when, within a growing future control of meaning at the level of the times, aggreformism becomes part of our heuristics. And so, this last chapter of this book, let alone this section, certainly is no place to attempt to develop a heuristics of the human organism.[75] Still, a symbolism will be introduced that will at least be continuous with efforts of earlier chapters, with the understanding that follow-up will be a community effort.

The organism *columbidae* may find crumbs of bread to eat on a piece of paper that has fallen to the ground near a café table. But, on that paper may be writings with which, for example, a letter-writer, a poet, or perhaps a mathematician has been occupied for days. Now, all that can be offered here is the thinnest of description. But, within preliminary self-attention, it is at least in keeping with heuristics already developed to suggest that human sciences will need to make progress toward identifying patterns of human sensibility that are not-merely-zoological,[76] that are merely coincidental with respect to zoological laws and principles, zoologically-explanatory of hunger, thirst, predation, avoidance, pleasure and pain.

---

[74] Lonergan, *CWL7*, 151.

[75] See, however, notes 72 and 84.

[76] "(T)he animal pertains to an explanatory genus beyond that of plant; that explanatory genus turns on sensibility; its specific differences are differences in sensibility; ...possessing a degree of freedom that is limited, but not controlled, by underlying materials and outer circumstances" (*CWL3*, 291).

Lonergan's compact description provides a lead:

> content of images provides the materials of mathematical understanding and thought; ... content of sensible data provides the materials of empirical method.[77]

And later:

> With the development of intelligence the reader already possesses some familiarity. The lower, otherwise coincidental manifold is provided by sensible presentations and imaginative representations.[78]

But, images and sensible presentations partly are neurological and zoological! While not yet working within a needed control of meaning, contemporary neuroscience and psychology are making considerable advances in identifying patterned combinations of layerings of aggregates of neurological and biochemical events corresponding, respectively, not only with moods and feelings, but with modes of speech, types of direct inquiry, insight, judgment, deliberation and decision.[79] Extending symbolisms already developed, a partial heuristics for human organisms capable of knowing,[80] deciding[81] and speaking[82] will be of the form

---

[77] *CWL3*, 291.

[78] *CWL* 492ff.

[79] There is now an extensive and growing body of results on neural correlates of: different feelings, different types of inquiry, thinking and deciding. There is a biophysics of brain activity identified through *functional magnetic resonance imaging* (fMRI). There is also ongoing progress in biochemistry and neuroscience of components of the brain. See, for example, B. Kraft, B. Gulyás and E. Pöppel (eds.), *Neural Correlates of Thinking*. Heidelberg: Springer, 2009. The challenge of assembling contemporary results within a verifiable heuristics is discussed in: Robert Henman, "Can brain scanning and imaging techniques contribute to a theory of thinking?," *Dialogues in Philosophy and Mental Neuro Sciences*, vol. 6, issue 2 (2013): 49-56; and Robert Henman, "Generalized Empirical Method: A Context for a Discussion of Language Usage in Neuroscience," *Dialogues in Philosophy, Mental and Neuro Sciences*, vol. 8, issue 1 (June 2015): 49-56.

[80] Bernard Lonergan, *Phenomenology and Logic*, vol. 18 in *Collected Works of Bernard Lonergan*, ed. Philip McShane (Toronto: University of Toronto Press, 2001), Appendix A, 319-323. Henceforth *CWL18*.

[81] Bernard Lonergan, *CWL18*, Appendix A, 319-323.

[82] "Spoken words are sounds with meaning: as sounds they are produced in the respiratory tract; ..." (Bernard Lonergan, *Verbum: Word and Idea in Aquinas*, vol. 2 of *Collected*

$$W_1(p_i; c_j; b_k; z_l; u_m).^{83}$$

Already, we can anticipate some advantages of having this kind of symbolism. Among other things, there will be a

> controlling power of W_1, either in initial or advanced work. Take the words 'phantasm' and 'consciousness.' The control alerts one, beginner or expert, to the reality referred to as zoological: a layered reality of physics and chemistry and botany. Without being thus alerted one could be stuck with a dangerous initial descriptive meaning, and with that meaning there is little chance of a broadened base of dialogue with modern searching.[84]

5.5 Intending History

In chapters two, three and four, a pre-heuristics was developed that calls for explanatory differentiation of things. Generically, physical things have

---

Works of Bernard Lonergan, eds. Frederick E. Crowe and Robert M. Doran (Toronto: University of Toronto Press, 1997); 14).

[83] See, Philip McShane, *A Brief History of Tongue*, Axial Publishing (1998), ch. 4, 122. The symbolism in the present book is incomplete, needing some further symbol such as 'q': "There is the possibility of new conjugate forms in man's intellect, will and sensitivity" (*CWL3*, 718ff). "There is our 'openness'" (*A Brief History of Tongue*, 122). But, that further 'q' layering, at least explicitly, goes beyond the present context.

[84] Philip McShane, "The Importance of Rescuing *Insight*," in *The Importance of Insight, Essays in Honour of Michael Vertin*, eds. John J. Liptay Jr. and David S. Liptay (Toronto: University of Toronto Press, 2007): 202. As developing human sciences increasingly reveal, "(o)rganic, psychic, and intellectual development are not three interdependent processes. They are interlocked, with the intellectual providing a higher integration of the psychic and the psychic providing a higher integration of the organic. Each level involves its own laws, its flexible circle of schemes of recurrence (see Section 5.5), its interlocked set of conjugate forms. Each set of forms stands in an emergent correspondence with otherwise coincidental manifolds on the lower levels. Hence, a single human action can involve a series of components: physical, chemical, organic, neural, psychic, and intellectual; and the several components occur in accord with the laws and realized schemes of their appropriate levels" (*CWL3*, 494ff; sec. 15.7.4, "Human Development"). Recall, also, the rule of explanatory formulation, noted in Section 4.4, above: "It is a rule of extreme importance, for the failure to observe it results in the substitution of pseudometaphysical mythmaking for scientific inquiry. ... (T)he absence of at least a virtual transposition from the descriptive to the explanatory commonly is accompanied by counterpositions on reality, knowledge, and objectivity" (*CWL3*, 528-9).

conjugate forms p_i; chemical things have aggreformically layered conjugate forms (p_i; c_j), and so on. The heuristics is neither *a priori* nor rigid. It is what science and philosophy of science have been discovering. It acknowledges subtleties of middle-types, such as viruses (relatively autonomic forms), or whatever other life-forms might be discovered within developing sciences. As the history of the sciences (including human sciences) reveals, at least five main layerings do not seem to be in doubt: There are physics; chemistry; botany; and zoology; and human sciences is about about organisms that are physical; chemical; botanical; zoological; and intellectual.[85]

How, though, are properties discovered and verified? For *columbidae*, the ability to see is verified when a living bird is seeing; the capacity to breathe oxygen is verified when a living bird breathes oxygen; a bird's abilities to metabolize food stuffs are verified when food stuffs are metabolized; and so on. That is, organisms are known by how they live and function in their environments of other living and non-living things, from *in natura* to *in laboratorium*.

But, what, then, is an organism's environment? It is an empirical question being explored by developing sciences. Think, say, of a pigeon's sensitivity to sunlight; the photochemistry of the earth's atmosphere; the chemical activity of the radiant sun, all joined through vast aggregates of conjugate acts[86] that are far too numerous to count. Continuing with elementary observations, note also that while a bird drinks water, water is part of a global water cycle. There is also a global nitrogen cycle; a global carbon cycle; and global oxygen cycle. And, there are many other global chemical cycles.[87] Through photosynthesis, a plant converts carbon

---

[85] See, however, note 83.

[86] See *CWL3*, 413; and the further development in basic position indicated by: "So it comes about ..." *CWL3*, 537. The quotation is given in Section 4.7 of this book.

[87] For a glimpse into the great complexities here, see, for example, Ei-Ichiro Ochiai, *Bio-inorganic Chemistry: A Survey* (Amsterdam: Academic Press, 2008), ch. 1, "The Distribution of Elements": 1.1 The Distribution of the Elements in the Earth's Crust, Seawater and Organisms; 1.2. The Engines that Drive the Biochemical Cycling of the Elements; 1.3. Flow of the Elements – The Biochemical Cycling; 1.4. Historical Change in the Biochemical Cycling of the Elements. See, in particular: Figure 1.5. The life process on Earth, and the movement of elements associated with life (p. 42); Figure 1.6. An outline

dioxide and water to glucose and oxygen. The (simplified) balanced chemical equation is:

$$Sunlight^{88} + 6CO_2 + 6H_2O \rightarrow C_6H_{12}O_6 + 6O_2.$$

In respiration (often called the opposite to photosynthesis), glucose and oxygen yield carbon dioxide and water. The (simplified) balanced equation for respiration is:

$$C_6H_{12}O_6 + 6O_2 \rightarrow 6CO_2 + 6H_2O.$$

But, animals and plants depend not only on water, oxygen and minerals, but also on other animals and plants - for example, for food, shelter, and the production of oxygen. Herons eat frogs; frogs eat flies; flies eat other insects, plants and fungi; and all of these can be found living together in wetlands. Flowering plants depend on animals for pollination, for distributing seeds, for the production of carbon dioxide. There are magnificently complex combinations of mutual dependencies, patterns of predator and prey, birth rates and death rates, almost countless symbiotic relationships among living and non-living entities in the biosphere and beyond. Within a basic position,[89] we find, then, that an organism's environment extends globally, but also includes the moon, the sun and, indeed, the entire cosmos. How might we extend our preliminary heuristics of aggreformic entities to embrace all verifiable mutual dependencies and functionings?

Recall from reflections in Chapter 2 that mutual dependencies are verified in fragmentary fashion, often remote to individual organisms *in vivo* or *in situ*. But, even when verified *in vivo* or *in situ*, indices of cycles and pathways of biochemical properties are not fully explanatory of the aggreformic activity of *E. coli*, or the growth and life of multi-cellular

---

of biogeochemical cycling of elements (p. 43); and Figure 1.7. The biogeochemical cycling of element carbon (p. 44).

[88] Recall similar issues in Chapter 2 on the TCA cycle. See previous paragraph. In the equation, *'sunlight'* is to be understood in terms of conjugate events and an organism's conjugate capacities-to-perform. There are, in particular, physical and chemical conjugate acts also verified in the aggregate solar mass.

[89] See *CWL3*, 413; and the further development in basic position indicated by: "So it comes about …" *CWL3*, 537. The quotation is given in Section 4.7 of this book.

columbidae. Nevertheless, biochemical equations such as the component equations of the TCA cycle are verifiable, and are part of ongoing advances in explanatory understanding of aerobic and non-aerobic metabolisms. A key insight here is that, for all of the complexities in mutual dependencies among entities of all kinds, there is a commonality, a commonality discovered by attending to how such dependencies are actually discovered and verified: That is, properties are discovered and verified within verifiable schemes of recurrence.

A *scheme of recurrence* is:

> a series of events *A*, *B*, *C*, … so related that the fulfillment of the conditions for each would be the occurrence of the others. Schematically, then, the scheme might be represented by the series of conditionals: If *A* occurs, *B* will occur; if *B* occurs, *C* will occur; if *C* occurs, …, *A* will recur. Such a circular arrangement may be involve any number of terms, and the possibility of alternative routes, and in general any degree of complexity.[90]

What, then, of the totality of schemes functioning together? There are verifiable scheme meshworks, a dynamic flexible totality of overlapping and mutually dependent schemes of recurrence. Within periods of relative stability, conditional probabilities $p = Prob \ (Y \ given \ X)$ remain bounded within subintervals $[a,b]$ of the unit interval: $0 < a \leq p \leq b \leq 1$. Ongoing developments in environmental science and ecology increasingly reveal complexities, diversities and scheme-works of sensitivities within even small portions[91] of the total scheme meshwork.

Following up from Section 5.4 above, what symbolism might help us name the totality? We already have a symbolism for aggreformic entities: $(p_i; c_j; b_k; z_l; u_m)$, where the internal symbols are present, or not, depending on genus and species. For example, $(p_i; 0; 0; 0; 0)$ is for physical things;

---

[90] *CWL3*, 141.

[91] Topsoil "provides a medium in which an astounding variety (and great numbers) of organisms live" (Richard Bardgett, *The Biology of Soil: A Community and Ecosystem Approach (Biology of Habitats Series)*, 1st ed. (Oxford: Oxford University Press, 2005), 2. In addition to organisms visible to the naked eye (such as earthworms, sowbugs, mites, centipedes, millipedes and spiders), in just ¼ teaspoon of healthy topsoil, numbers of microscopic organism are of order 50 nematodes; 60,000 algae; 70,000 amoebae; 2,900,000 actinomycetes; and 25,000,000 bacteria.

(p_i; c_j; 0; 0; 0) is for chemical things; (p_i; c_j; b_k; 0; 0) is for plants; and so on. But, things function within the totality, in and across aggreformic layerings. So, we need another symbol, 'f' say ('f' for 'functionings'). There are (internal) relations,[92] for example, potency: form: act; and other real relations R to be discovered.[93] And, as already discussed briefly, there are probabilities P. The totality is a sum 'S' of all things. But, what kind of sum? Schemes of recurrence are verifiable. Altogether, then, a preliminary symbolism for the totality is

$$S_{RP}f(p_i;\ c_j;\ b_k;\ z_l;\ u_m),^{94}$$

where the S includes all actual schemes of recurrence of all things.

This is an extremely compact notation. At the same time, it is empirically grounded, and emergent from developments in the contemporary sciences. As a heuristics, it includes, for example, complex substructures that are the biochemistry of the TCA cycle of each aerobic organism. Likewise, it includes schemes of recurrence being discovered in particle physics, astrophysics and gravitation theories. It includes the vast complexes of schemes of recurrence of all actual population dynamics that involve, and cross through, aggreformic layerings of aggreformic entities, terrestrial and beyond. Probabilities 'P' are actual - Normal, Poisson, Continuous, Discrete, Mixed, whatever the case may be. What is in evidence, then, is that the biosphere is part of a vast (approximately-) Markov process.[95]

---

[92] *CWL3*, sec. 16.2.

[93] *CWL3*, sec. 16.2.

[94] For further discussion on this, see Philip McShane, *A Brief History of Tongue* (Axial Publishing, 1998), 120-121. Note that the 'u_m' component will be discussed in the Epilogue.

[95] To understand Markov processes, one needs to climb into modern statistical analysis and probability theories. For readers with some background in probability theory and mathematical analysis, but not yet familiar with Markov theory, one may consult one of the many available introductions. See, for example, Howard M. Taylor and Samuel Karlin, *An Introduction to Stochastic Modeling*, 3rd ed., San Diego: Academic Press, 1998. One needs to be alert to errors of the types already discussed, wherein probable judgment is confused with empirical probability, and probability distributions (ideal relative frequencies) mistakenly are applied to individual events and occurrences. This is not to suggest that there is no such things as *reasonable betting*. See notes in Section 1.8.

However, while some regions of space and time have had long periods of relative stability, the universe is known to be dynamic. Not only have there been various cosmological epochs, but, in the Sun Formation epoch, the biosphere and its ecosystems have been changing in fundamental ways. Recently, too, there is a global warming, involving major changes in world biogeography mainly brought about by human activity.[96] These changes are being seen at all levels: in atmospheric physics, ocean thermodynamics and ocean chemistry, global weather patterns, and world ecosystems. When an organism can no longer survive, schemes in which it previously participated change, or in some cases, are eliminated. And, contemporary environmental science anticipates ongoing changes: atmospheric and geological, animal habitats, extinctions of species, emergence of new species, and so on.[97] In short, the totality $S_{RP}f(p_i; c_j; b_k; z_l; u_m)$ has been changing and is changing. In our symbolism, therefore, we need another symbol, 'H' say, for intending History as emergent fact: $HS_{RP}f(p_i; c_j; b_k; z_l; u_m)$. World process, then, is not any single global Markov process, but a time-ordered series of emergent (approximately-) Markov processes.

5.6 Emergent Probability

5.6.1 *Probabilities of schemes of recurrence*

Empirical probability of events and occurrences was discussed briefly in Section 1.7. And, in contemporary evolutionary studies, there are statistical results on empirical probabilities of emergence of species.[98] But, if we look to not just events and occurrences, but also to schemes of

---

[96] Charles F. Keller, "Global warming: a review of this mostly settled issue," *Stochastic Environmental Research and Risk Assessment*, vol. 23, no. 5 (July, 2009): 643-676.

[97] Scheme stability is crucial in ecology and environmental science. What are the ranges of circumstances under which a scheme will continue to function? Or, under which some modified version of the scheme will continue to function? As contemporary environmental science has been discovering, some schemes are relatively robust, while others are significantly altered, or eliminated, when there are even minor changes to an ecosystem.

[98] See, for example, the methods used in, S. F. Elena and R. E. Lenski, "Evolution experiments with microorganisms: the dynamics and genetic bases of adaptation," *Nature Reviews Genetics*, 4, (2003): 457-469.

recurrence, further questions can arise. As mentioned above in Section 5.5, schemes of recurrence have come and gone through various eras of life on earth. In some cases, schemes are enduring, in other cases, less so.[99] Are there also, then, empirical probabilities of schemes of recurrence, of their emergence, and of their survival?

We might well begin by looking to some of the familiar chemical cycles, such as the Nitrogen cycle, Carbon cycle, or Oxygen cycle, which it would seem co-evolved in the Proterozoic Oceans.[100] Detailed investigation into their emergence would take us into the great world of contemporary biochemistry, with a special focus in the now-established sub-field called paleobiology. In the case of the nitrogen cycle, for example, it is now known that the modern nitrogen cycle was attained in stages, gradually, first through various prebiotic nitrogen cycles, followed later by the emergence of biotic cycles, both anaerobic and aerobic.

> Atmospheric reactions and slow geological processes controlled Earth's earliest nitrogen cycle, and by ~2.7 billion years ago, a linked suite of microbial processes evolved to form the modern nitrogen cycle with robust natural feedbacks and controls.[101]

So, while the nitrogen, carbon and oxygen cycles are well known, an investigation into probabilities of (their stages of) emergence, as well as of their present-day functionings, would need all of the sophistications of contemporary biochemistry, biophysics,[102] including contemporary paleo-biochemistry and contemporary paleo-biophysics.

What might we do here, to make a start toward answering the question about probabilities of emergence of schemes of recurrence? Instead of looking to the more familiar Nitrogen, Carbon and Oxygen cycles, let's

---

[99] See note 97.

[100] Katja Fennel, Mick Follows and Paul G. Falkowski, "The Co-evolution of the Nitrogen, Carbon and Oxygen Cycles in the Proterozoic Ocean," *American Journal of Science*, vol. 305 (June - Oct., 2005): 526-545.

[101] Donald E. Canfield, Alexander N. Glazer and Paul G. Falkowski, "The Evolution and Future of Earth's Nitrogen Cycle," *Science*, vol. 330, (2010): 192-196. http://www.sciencemag.org/content/330/6001/192.

[102] Paul G. Falkowski and Linda V. Godfrey, "Electrons, life and the evolution of Earth's oxygen cycle," *Philosophical Transactions of the Royal Society B,* 363 (2008): 2705–2716.

look again to the TCA cycle. The sophisticated TCA cycle was relatively late to emerge in the biosphere, and investigating its emergence certainly also would call on the full reaches of contemporary biochemistry and paleobiochemistry. However, thanks to having already discussed the TCA cycle in earlier chapters, we can temporarily avoid having to open up a completely new discussion. As a continuation of earlier discussion, and looking to some of what is known about the origins of TCA cycle, we can get some inkling of the real possibility of, but also challenges and subtleties that would be involved in, discovering probabilities of emergence of schemes of recurrence.

On present understanding, there are inter-dependencies among what often are called "intermediates" of the citric acid cycle. In aerobic organisms, if an intermediate of the TCA cycle is found to be present (aggreformically) in concentrations, then in a short time each of the other intermediates also will be found (at relative rates and concentrations that in many cases can be estimated from reaction equations *in vitro* and *ex vivo*).

There was, however, a time in prehistory when there were was no TCA cycle,[103] when anaerobic microorganisms dominated world ecosystems. Eventually, though, things began to change.

> Several distinct, incomplete cycles reflect adaptations to different environments. Their distribution over the phylogenetic tree hints at precursors in the evolution of the citric-acid cycle.[104]

Emergence of the TCA cycle and its various precursors depended on

---

[103] There was an age "Before Life Began," Futuyma, ch. 5, 102-126.

[104] Oliver Ebenhöh and Reinhart Heinrich, "Evolutionary Optimization of Metabolic Pathways. Theoretical Reconstruction of the Stoichiometry of ATP and NADH Producing Systems," *Bulletin of Mathematical Biology*, vol. 63, issue 1 (January 2001): 21-55. Martin A. Huynen, Thomas Dandekar and Peer Bork, "Variation and Evolution of the Citric Acid Cycle - A Genomic Perspective", *Trends in Microbiology*, vol. 7., no 7. (July 1999): 281-291. See also Enrique Melendez-Hevia, Thomas G. Waddell, Marta Cascante, "The Puzzle of the Krebs Citric Acid Cycle: Assembling the Pieces of Chemically Feasible Reactions, and Opportunism in the Design of Metabolic Pathways During Evolution," *Journal of Molecular Evolution* 43 (1996): 293–303.

sufficient stability and robustness of prior schemes.[105] In other words, the TCA cycle was not attained all at once, but in increments and in different ways for different genera and species of one-celled organism[106] surviving in diverse ecosystems. Remember that empirical probabilities are ideal relative frequencies. So, in order to enquire into probabilities of emergence of the complete citric acid cycle, we will need representative samples of that emergence. And to do that, we need to look to large enough periods of time and large enough numbers within the geo-evolutionary phylogenetic tree that includes precursors to the cycle in present-day organisms.

What were the relative frequencies of emergence in organisms, of various steps of the present day TCA cycle? Determining those fractions is a question for biochemical evolutionary studies. However, such studies already have discovered that over large periods of time and large regions in the biosphere, intermediates of the TCA cycle are not *statistically independent*.[107] In diverse environments, populations of micro-organisms

---

[105] This also raises question of probabilities of survival of schemes. See note 113. Regarding the TCA cycle itself, there is an emerging literature on not only its generalized robustness, but, among biochemically feasible metabolic pathways, its special efficiency and energy economy. Enrique Melendez-Hevia, Thomas G. Waddell, Marta Cascante, "The Puzzle of the Krebs Citric Acid Cycle: Assembling the Pieces of Chemically Feasible Reactions, and Opportunism in the Design of Metabolic Pathways During Evolution," *Journal of Molecular Evolution* 43 (1996): 293–303. See also, Martin A. Huynen et al., "Variation and Evolution of the Citric Acid Cycle - A Genomic Perspective."

[106] For a sampling from the literature, see the following articles: A.H. Romano and T. Conway, "Evolution of carbohydrate metabolic pathways," *Research in Microbiology*, vol. 147, issue 6-7 (Jan., 1996): 448-455; Claus Schnarrenberger and William Martin, "Evolution of the enzymes of the citric acid cycle and the glyoxylate cycle of higher plants. A case study of endosymbiotic gene transfer," *European Journal of Biochemistry*, vol. 269, (2002): 868-883; Miklós Péter Kalapos, "The energetics of the reductive citric acid cycle in the pyrite-pulled surface metabolism in the early stage of evolution," *Journal of Theoretical Biology*, vol. 248 (2007): 251–258; and David S. Ross, "The Viability of a Nonenzymatic Reductive Citric Acid Cycle - Kinetics and Thermochemistry," *Origins of Life and Evolution of Biospheres* (2007) vol. 37 (2007): 61–65.

[107] See references in note 114. Differently defined events *A* and *B* are said to be *statistically independent* if relative actual frequencies of *A* are not statistically correlated with the relative actual frequencies of *B*. In the biosphere, given large enough time and large enough numbers, there is in fact little evidence that there are statistically independent events. The

functioned through various precursors of the citric acid cycle, eventually reaching a cusp of some next shift. What, then, is the probability of emergence of the complete TCA cycle? Because intermediates are not statistically independent, probabilities of emergence are approximated by a sum of probabilities of intermediates found in precursors!

The reflection here has been an (overly) brief excursion into the problem of working out probabilities of emergence of schemes. Still, let us now take a moment to look to Lonergan's pointers on the problem. For, in view of empirical results known about the emergence of the TCA cycle (and if one gathers what is known about the evolution of the Nitrogen, Carbon and Oxygen cycles[108]), the description below will be seen to provide a basis for future empirical investigation:

> Let us suppose that the set of events *A, B, C,* … satisfies a conditioned scheme of recurrence, say *K*, in a world situation in which the scheme *K* is not functioning, but in virtue of the fulfillment of prior conditions could begin to function. … In brief, if any events of the scheme (*K*) were to occur, then, other things being equal, the rest

---

biosphere is sensitive to even small changes in sustained relative frequencies. However, for sufficiently short time periods, the empirical probability of a pair (*A* and *B*) often can be approximated by a product of conditional probabilities for *A* and *B*, respectively. Familiar examples come from games of chance. One may think of, say, two differently weighted dice. Suppose that they are both unfairly weighted, so that for the first die, the empirical probability of '1' is 3/5; and for the second, the empirical probability of '1' is 1/2. What, then, is the empirical probability for the outcome that is 1 for the first die; followed by 1 also for the second die? One way to begin sorting this out is to make a 6 by 6 table of all possible outcomes of the form (result of first die, result of second die). There are 36 possible ordered pairs: (1, 1), (1, 2), …, (2, 1), (2, 2), …, (6, 5), (6, 6). If we suppose representative samples, then (with a bit of work) one finds that the ideal relative frequency of "snake eyes" (1, 1) probably is approximated by $(3/5) \times (1/2)$. As with any empirical probability, one needs to verify the result in representative samples. Similar, but far more complex, problems were "posed to Pascal by a highly intelligent gentleman addicted to gaming. …The necessary mathematics all developed from the fundamental principles of mathematical probability laid down by Fermat and Pascal in about three months of painstaking application of uncommon sense" (Eric Temple Bell, *The Development of Mathematics* (New York: McGraw-Hill, 1945), 155). Again, one needs to distinguish the mathematics (of permutations, combinations, distributions and measure theories) from empirically verifiable probabilities. See also "games of chance" in *CWL3*, 85, and notes in Section 1.8 above.

[108] See notes 100, 101 and 102.

of the events in the set would follow.

In this case we may suppose the probabilities of single events are the same as before (p for A, q for B, ...), but we cannot suppose that the probability of the combination of all events in the set is the same as before. As is easily to be seen,[109] the concrete possibility of a scheme beginning to function shifts the probability of the combination from the product pqr ...[110] to the sum $p + q + r + ...$

Now a sum of a set of proper fractions[111] $p, q, r, ...$ is always greater than the product of the same fractions.[112] ... It follows that, when the prior conditions for the functioning of the scheme of recurrence are satisfied, then the probability of the combination of events constitutive of the scheme leaps from a product of a fractions to a sum of fractions.[113]

5.6.2 *Emergent probability and Markov processes*

As already mentioned in sections 5.1 and 5.2, there is what is called the Solar System Formation epoch, which includes all periods of life on earth so far. A broad division of the "the grand history of life"[114] on earth is into four eras: Precambrian, Paleozoic, Mesozoic and Cenozoic. Each of these eras has its eons and epochs.

---

[109] What Lonergan meant is a question for future interpretations. However, one already can marvel at his genius: Ongoing contemporary searchings and confusions about emergence (see, e.g., secs. 5.2 and 5.3) reveal that generally it is *not* easily to be seen.

[110] See note 107.

[111] *Proper fraction*: The positive integer numerator is strictly less than the positive integer denominator. A proper fraction is, therefore, strictly less than unity.

[112] For example, if there are two fractions, $(2/3)$ and $(1/4)$, then $(2/3)(1/4)$ is less than $(1/4)$. In the same way, $(1/4)(2/3)$ is less than $(2/3)$. In other words, $(2/3)(1/4)$ is less than both $(2/3)$ and $(1/4)$; and both of these are less than the sum $(2/3) + (1/4)$. The general case is easily established. In the case of two probabilities, note that for any $0 < p, q < 1$, we have $pq < \max(p, q) < p + q$.

[113] *CWL3*, 143-144. Compare this with probabilities of pairs of independent events. Note that the probability of emergence of a scheme is distinct from the probability of its survival. While schemes may be more, or less, stable under changing circumstances, there is always the proviso, 'other things being equal.' The probability of survival a scheme is "the probability of the nonoccurrence of any of the events that would disrupt the scheme" (*CWL3*, 144).

[114] Futuyma, ch. 5, 101.

The early atmosphere had little oxygen, so the earliest organisms were anaerobic.[115]

However,

> as oxygen built up in the atmosphere, many organisms evolved the capacity for aerobic respiration, as well as mechanisms to protect the organism against oxidation.[116]

And, in present times, there are unknown multitudes[117] of aerobic organisms whose metabolism depends on the energy efficient TCA cycle[118] in a now oxygen-rich atmosphere.

How can we hold all of this together? From section 5.5, we have a preliminary heuristics, $HS_{RP}f(p_i; c_j; b_k; z_l; u_m)$, for intending history as emergent fact. In view of section 5.6, our heuristics needs to include series of probabilities of emergence of species and of schemes of recurrence, as well as probabilities of stabilities, survivals and extinctions. The emergence of chemical schemes of recurrence depended on the prior functioning of physical schemes; the emergence of schemes of recurrence of anaerobic and eventually aerobic organisms depended on the prior functioning of chemical schemes; and so on. Verifiably, the totality is a

> successive realization in accord with successive schedules of probability of a conditioned series of schemes of recurrence,[119]

what we can call an *emergent probability*.[120] Still, while naming and describing a few aspects of the totality is a beginning, we are far from reaching an explanatory heuristics of "the immanent form or

---

[115] Futuyma, 102.

[116] Futuyma, 102.

[117] Recent estimates of earth's biodiversity are given in: Andrew Pullin, *Conservation Biology. Cambridge*: Cambridge University Press, 2002. See also, Camilo Mora et al., "How Many Species Are There on Earth and in the Ocean?", *Public Library of Science (PLoS), Biology*, 9(8) (August, 2011): e1001127. For an indication of actual numbers or organisms, see, for example, note 91.

[118] See note 105.

[119] *CWL3*, 148.

[120] *CWL3*, 145. See secs. 4.2.4 (144-148) and 4.2.5 (148-151).

intelligibility."[121] The work of the previous chapters, though, does begin to bring out a main feature of total process: Total process is a vast time-ordered series of approximately-Markov processes.[122]

5.7 The Problem of Implementation

A main purpose of this book has been to make progress in discovering the possibility and desirability of a generalized empirical method. The balanced method was named and briefly described by Lonergan in his book *Insight*.[123] There, and throughout his *opera omnia*, we find discussions about the need to move toward the new method. As already noted in the Preface, in one of his later articles Lonergan provided a more precise definition of *generalized empirical method*.[124]

The need is increasingly evident but, so far in history, generalized empirical method has not yet been tried. As mentioned in Section P.4 of the Preface, some individuals have been discussing Lonergan's leading ideas, in the foundations of science as well as in other areas. A method, though, is a community achievement. And, so far, generalized empirical method is not yet operative and influential in the world scientific and philosophic communities. Contemporary methods tend to be object-oriented and, in particular, do not recognize the need for a control of meaning that can be attained through self-attention. From Kindergarten on, young minds are schooled into being self-screened. Contemporary philosophy of science does not generally require that views about science be verifiable in scholars' experience in scientific practice. Instead, emphasis is on conceptual and imaginary constructs, and logical analysis, rather than on realities to be known through experience.[125] It is not that

---

[121] *CWL3*, 195. See also, Section 5.7, and Epilogue.

[122] In a Markov process, states and probabilities are fixed. I include the word "approx.-imately" to heuristically point to the fact that even during periods of general stability, events, occurrences and their probabilities tend vary in time. What events, occurrences and probabilities are during any time interval is an empirical problem.

[123] The first edition of the book was: Bernard Lonergan, *Insight: A Study of Human Understanding*. London: Longman, Green and Co., 1957. In the 1st ed., see pp. 72ff.

[124] Bernard Lonergan, *A Third Collection*, 141.

[125] See, *basic position*, *CWL* 413.

there are not brilliant, creative and insightful scholars. But, contemporary philosophic methods allow for more or less endless debate. And so one of the consequences is that potentially important insights cannot gain traction in ways that might lead to cumulative and progressive results.

Again, this is not merely an academic matter. Indeed, confusions in science and philosophy of science contribute to, and also emerge from, confusions in general education. Education is influenced by textbook industries and establishment economics. Establishment economics combines the mathematics of gambling[126] with institutionalized greed. The resulting cultural and ecological damage being propagated in the name of the "insanity"[127] called establishment economics is too well-known to mention here in any detail. In financial quarters, there has been some growing interest in environmental problems and reducing Global Warming.[128] But, there has been little openness or interest in Economics

---

[126] For one example, there is the Black and Scholes (and Merton) equation for determining the values of derivatives. In 1997, Merton and Scholes were awarded a Nobel Prize for this work.

[127] Roger Terry, *Economic Insanity: How Growth-Driven Capitalism Is Devouring the American Dream* (San Francisco: Berrett- Koehler: 1995). "It has become obvious that we must find an alternative to the soulless quest for profit that drives our market system. .... The conservative solution of setting the market free from government meddling only intensifies all the problems outlined above. Additionally, it intensifies the fragmenting of society into diverse economic classes and exacerbates our dependence on debt and welfare. The liberal solution of redistributing income, on the other hand, creates dependence and lack of initiative among the poor and breeds resentment and callousness among the wealthy. Neither of these solutions has more than a tenuous connection to reality" (Roger Terry, "Rethinking the profit motive," *World Business Academy Perspectives*, vol. 9, no. 4, (1995): 26).

[128] For instance, Henry Paulson and Robert Rubin (both former Treasury Secretaries for the USA, and executive officers with Goldman Sachs) were, for a time, promoting changes in economic policy and corporate practice that would better protect ecologies, and help re-stabilize weather patterns." CNN's Fareed Zakaria, GPS, features an interview with the former U.S. Treasury Secretary under George W. Bush, Henry Paulson, and the former U.S. Treasury Secretary under Bill Clinton, Robert Rubin. Paulson and Rubin speak with Fareed about their new report on the future of our environment if Americans do not start taking preventative measures against climate change, the cost of inaction, and the limitations to progress posed by Washington. Additionally, Paulson and Rubin also speak with Fareed about the U.S. fiscal outlook and economic recovery" (*CNN Press Room*,

Academic Departments to find and correct fundamental errors[129] in what obviously is a flawed establishment economics. And so it has been going: economists and

> philosophers for at least two centuries, through doctrines on politics, economics, education, and through ever further doctrines, have been trying to remake man and have done not a little to make life unlivable.[130]

It is an especially difficult and confusing time: Peoples clash; schooling and higher education systematically disorient young minds from their own dynamics of wonder, knowing and doing; global ecosystems are being severely damaged; and while it is not the whole story, to some extent, difficulties are cumulative.[131] We are participants in emergent probability, "executor(s) of the emergent probability of human affairs."[132] But, in present times, much of what we are doing is mutilating the very ground from which we emerge.

What then, of our part, in science and philosophy? If we continue with familiar methods, we can only expect already familiar kinds of results. But, if generalized empirical method is desirable, and ultimately is to be normative, what will implementation look like? And, for that matter, how in the world (process)[133] can we get there from here? As history shows,

> (t)here is nothing to prevent an intelligent (person) ... from discovering ... structures that they cannot escape, and from generalizing from the totality of examined instances to the totality of possible instances.[134]

---

http://cnnpressroom.blogs.cnn.com/2014/06/29/fmr-u-s-treasury-secy-rubin-on-climate-change-the-risk-here-is-catastrophic/).

[129] Present economic ills brought about through establishment economics have their roots in errors identified more than seventy years ago by Bernard Lonergan. See, for example, Bernard Lonergan, *For a New Political Economy, Collected Works of Bernard Lonergan*, vol. 21, ed. Philip McShane. Toronto: University of Toronto Press, 1998.

[130] Bernard Lonergan, *Topics in Education, Collected Works of Bernard Lonergan*, vol. 10, eds. Robert M. Doran and Frederick E. Crowe (Toronto: University of Toronto Press, 1993), 232.

[131] In *Insight*, Lonergan provides detailed description of present patterns of cumulative decline (*CWL3*, sec. 7.8, 251 - 267), and introduces the name "longer cycle of decline."

[132] *CWL3*, 252.

[133] See "world process," indexed in *Insight, CWL3*, 874.

[134] *CWL3*, 425.

But, as history also shows, and present times confirm, the odds against that happening are high. At present, probabilities of such growth are near zero. Evidently, making the transition to the balanced empirical method will be a massive and subtle enterprise, a global shifting of cultures and education out of deeply established disorientations. What might we do to lift empirical probabilities of emergence of the new balanced empirical method from near zero to some epsilon effectively greater than zero?

Throughout *Insight*, Lonergan was concerned with the problem of implementation.[135] In Chapter 7 of that book, he gave a name to what, at the time,[136] was an as yet unknown X, a solution to the problem of cumulative cultural decline, a solution that would be an actually attainable and sustainable *cosmopolis*.[137] He was, though, able to provide "a series of notes"[138] and to indicate "a few aspects."[139] It was later, in February of 1965, that he made his major breakthrough[140] to a *Practical View of History*,[141] a discovery that he called *functional specialization*.[142] An Epilogue is needed, however, to briefly discuss these further issues.

---

[135] In Chapter 14 of *Insight*, he identifies metaphysics with "the conception, affirmation and implementation of the integral heuristics structure of proportionate being" (*CWL3*, 416).

[136] "So far from solving it in this chapter, we do not hope to reach a full solution in this volume" (*CWL3*, 267).

[137] *CWL3*, 263.

[138] *CWL3*, 259.

[139] *CWL3*, 263.

[140] The discovery document now is available from a number of sources: The Bernard Lonergan Archive, http://www.bernardlonergan.com/index.php; 47200D0E060 / A472 V\7\1 - Functional specialties: Breakthrough page; Lonergan Archives, Lonergan Research Institute, Regis College, Toronto, Canada, http://www.lonergan-lri.ca/, Batch V.7.a; or Pierre Lambert and Philip McShane, *Bernard Lonergan – His Life and Leading Ideas* (Vancouver: Axial Publishing, 2010), 160.

[141] Darlene Mary O'Leary, *Lonergan's Practical View of History*, Master of Arts in Theology, University of St. Michael's College, Toronto, Canada, 1999. Hardcopy is available at the John M. Kelly Library, University of St. Michael's College, Toronto; as well as the library at Regis College, Toronto.

[142] Bernard Lonergan, "Functional Specialties in Theology," *Gregorianum*, vol. 51 (1970): 537-540. See also ch. 5 in, Bernard Lonergan, *Method in Theology*. London: Darton, Longman and Todd, 1972.

Epilogue

What is Science?

Abstract: The focus of the book has been on method in the sciences. But, there is another part of the problem that, so far, has not been addressed: 'What is Science?' This Epilogue relates the question, 'What is Science?'; a discovery of Bernard Lonergan that he called *functional specialization*[1]; and the problem of implementation discussed briefly in Section 5.7. I am, of course, not going to offer an answer to the question 'What is Science?', and so the question mark in the title of the chapter is appropriate. However, taking help from Lonergan's later discovery called *functional specialization*, we can anticipate that mature science will be an eightfold collaboration.

E.1 A Geo-Historical Compound Enterprise

What is science? Over the centuries, answers to the question have been provided from within various traditions in, for example, the West, China, Africa and India.[2] In the West,

> (t)here has been … a move towards what we might call clarity or ideal typology such that deviants from that clear type were considered, in some way, inferior. This is true whether one considers formulations of paradigms, such as constitute texts on philosophy of science, or discussions of paradigm-shifts, such as occur in the

---

[1] See note 19.

[2] See, for example, Helaine Selin, ed., *Encyclopaedia of the History of Science, Technology, and Medicine in Non-Western Cultures*, 2 vol. set, 2nd ed. Dordrecth: Kluwer, 2008.

Kuhnian tradition.[3]

Instead, here, let's see what preliminary observations can be made about scientific practice. Recall, for instance, quotations at the beginning of the Preface. They describe an evident continuity of science and philosophy of science. But those same statements implicitly acknowledge that there are distinctions needing identification. Chapters 1 to 5 of the book invites exploration of an empirical method by which, among other things, we eventually would be able to grow in being luminous about such distinctions - in conception, affirmation and implementation.[4] To be sure, such control of meaning will be a somewhat distant future achievement.[5] But, already it is widely recognized that scientific progress is, in fact, a compound enterprise; and that, whatever distinctions there are, progress in science and in philosophy of science are mutually dependent within the whole enterprise.

If, though, science is a compound enterprise, might we not attempt to be as complete as possible in identifying all mutually dependent parts of the enterprise? Or, are we to ignore some parts or some aspects of the process? What other kinds of work, then, contribute to, and are mutually dependent within scientific development? Notice here that the problem continues to be empirical, analogous to that of basic anatomy when describing organs, connectivities and other parts.[6] The empirical approach here asks that we identify all "organs" and "connectivities" found in scientific practice.

---

[3] See Introduction, in Robert Henman, "Implementing Generalized Empirical Method in Neuroscience by Functionally Ordering Tasks," *Dialogues in Philosophy, Mental and Neuroscience*, vol. 9, issue 1, June 2016. Regarding origins in the West, as quoted in Lonergan, *Method in Theology*, 3, and in Henman, "Implementing Generalized Empirical Method in Neuroscience," (2015): "Throughout the whole of (Aristotle's) works we find him taking the view that all other sciences than the mathematical have the name of science only by courtesy, since they are occupied with matters in which contingency plays a part" (Sir William David Ross, *Aristotle's Prior and Posterior Analytics* (Oxford: Clarendon Press, 1949), 14. See also, pp. 51 ff. A revised reprint is: Oxford: Clarendon Press, 2000.

[4] *CWL3*, 416.

[5] That achievement will be practical. See, *CWL3*, 416.

[6] See, for example, the anatomy book referenced in ch. 2: Robert B. Chiasson, *Laboratory Anatomy of the Pigeon*, 3rd Ed., Dubuque (Iowa): W. C. Brown, 1984.

At present, an empirical approach is not the norm.[7] Still, in some respects, beginnings toward an empirical approach are not entirely new. For instance, Heather Douglas observes that Thomas Kuhn's view was

anything but a picture of unified science, at least at the global level.[8]

And, the paper by Douglas goes on to shed light on a neglect of what has been called "applied science," a rhetorical distinction that emerged historically and that does not stand up to philosophical scrutiny.[9] Toward the end of her paper, Douglas suggests that

if scientists want to reject the narrow sense of scientific progress, … , they will have to accept a more socially, ethically mediated conception of progress, one that takes into account all of science, both pure and applied. To construct a sense of scientific progress that sounds genuinely like progress, with all its positive connotations, we are going to have to embed science even more fully in society.[10]

But, then there is the additional question: 'What are the *applied sciences*?' Lists of areas called *applied science* have been growing for decades, with publications filling libraries. To name just a few broad divisions, these include, but certainly are not limited to, emerging specialties in applied quantum-electronics, applied physical-chemistry, applied bio-chemistry, neurochemistry, interdisciplinary sciences, applied mathematics, computer sciences, engineering sciences, industrial sciences,

---

[7] See, for example, note 3.

[8] Heather Douglas, "Pure science and the problem of progress," *Studies in History and Philosophy of Science, Part A* 46 (2014), 55. See also, Till Grüne-Yanoff, "Introduction: Interdisciplinary model exchanges," *Studies in History and Philosophy of Science, Part A,* vol. 48 (2014): 52-59. Ernst Mayr also advocated the need for philosophy of science to look to scientific practice. See, for example, "Do Thomas Kuhn's scientific revolutions take place?", ch. 9 in Ernst Mayr (1904-2005), *What Makes Biology Unique? Considerations on the Autonomy of a Scientific Discipline* (Cambridge: Cambridge University Press, 2004). "There are numerous more or less independent aspects of Kuhn's thesis, but they cannot be discussed profitably without looking at concrete cases. It is necessary to study particular sciences at particular periods and ask whether theory change did or did not follow Kuhn's generalizations. I have therefore analyzed a number of major theory changes in biology" (Mayr, 160).

[9] Douglas, 55.

[10] Douglas, 63.

agriculture sciences, veterinary sciences, medical sciences,[11] applied environmental sciences, technologies and economics.

What we have, then, is that: (1) the body of scientific practice involves all of what traditionally are called pure sciences; applied sciences; and philosophy of science; and (2) all of these are mutually dependent in the scientific enterprise. But, are there, perhaps, still other mutually dependent areas of inquiry in the enterprise? The question itself partially reveals that there are. For, the question asks about what has been going on in the scientific enterprise. But, historical understanding is a distinct focus, and at the same time contributes to scientific progress. Recall too that, in Section 1.9, it was observed that (systematic) historical understanding eventually needs to be part of a control of meaning at the level of the times: for results are emergent in historical process. But, in the whole scientific community, there are still many other activities that implicitly or explicitly are mutually dependent within scientific practice. There are, for example, aesthetics of science, theories about possible developments, theories about language in the sciences, theories about scientific communications and education, community outreach, ongoing and emerging applications and technologies. The data on scientific practice and development is vast, emergent and geo-historical.

E.2 The Wheel of Progress

How might we follow up from these few and of course only preliminary observations? What more might we gather about science, about scientific progress, and about how science is embedded in, and part of, progress in society[12]? Or, is scientific progress merely whatever developments (and confusions) happen to be influential at a given time?

The questions might seem to point well beyond what we can answer at this time in history. Fortunately, however, we don't need to reinvent the

---

[11] For example, Steve Webb, "The contribution, history, impact and future of physics in medicine," *Acta Oncologica*, 48 (2009): 169-177. See also, Mark Jackson, ed., *The Oxford Handbook of the History of Medicine*. Oxford: Oxford University Press, 2011.

[12] Sal Resitvo, ed., *Science, Technology and Society: An Encyclopedia*. Oxford: Oxford University Press, 2005. See also note 10.

wheel.[13] Part of Lonergan's decades-long inquiries were given to investigating methods of natural sciences, human sciences and theology. At the beginning of *Method in Theology*, he speaks of the need to

appeal to the successful sciences to form a preliminary notion of method.[14]

But, he also points to the need for "a third way,"[15] the need to

go behind the procedures of the natural sciences to something both more general and more fundamental.[16]

Lonergan had a reaching[17] multi-category[18] control of meaning, and was able to discern the pre-emergence in history of a normative division of labor in the academy, a cyclic collaboration that he called *functional specialization*.[19] As named by Lonergan, eventually there will be eight functional specialties: functional research, functional interpretation, functional history, functional dialectics; and functional foundations, functional doctrines, functional systematics and functional communications. The functional specialties were "conceived along interdisciplinary lines"[20] and, in fact, bear out not only in the human and

---

[13] See note 44.

[14] Lonergan, *Method in Theology*, 4.

[15] Lonergan, *Method in Theology*, 4.

[16] Lonergan, *Method in Theology*, 4.

[17] While much of Lonergan's published work was in theology, his background included the natural sciences, 20th century mathematics, mathematical logic and physics up to the 1950's.

[18] See Lonergan, *Method in Theology*, secs. 11.5-11.8, 281-293. In particular, see 286-287; and note 2, page 7.

[19] Bernard Lonergan, "Functional Specialties in Theology," *Gregorianum*. 1970: 537-54. See also ch. 5 in, Bernard Lonergan, *Method in Theology*. London: Darton, Longman and Todd, 1972.

[20] Bernard Lonergan, *A Third Collection, Papers by Bernard J. F. Lonergan, S. J.* Ed. Frederick E. Crowe, S. J. (New York: Paulist Press, 1985), 113.

natural sciences,[21] but in all areas of academic endeavor.[22]

I have been suggesting that we attempt to be as complete as possible in searching out all types of work that are part of, contribute to and are dependent on the sciences. But, with the help of Lonergan's descriptions of eight types of task,[23] we can initiate new inquiries into the geo-historical scientific enterprise. I note that, in these new inquiries, there is no avoiding the need of self-attention. For, in speaking of eight tasks, it is not being suggested that an eightfold structuring be imposed on the sciences, or that that the eightfold division of labor is a typology in the old style of philosophy.[24] As mentioned in the paragraph above, the discovery is empirical, or, if one prefers, generalized empirical.

Within a mode of self-attention, and with the help of Lonergan's leads, we can look to scholarship, communications, collaborations, developments and applications attained and being attained within the scientific community. While there are ongoing differentiations and developments in all areas, it can become increasingly (self-) evident that eight main types of work are what have been, and are, driving the scientific enterprise in its many dimensions, including those that are part of what traditionally are called pure science; applied science; philosophy of science; and society.[25]

In various areas, preliminary studies already have been done toward investigating the eightfold dynamics discovered by Lonergan. These areas include: physics, housing science, philosophy, science and technology, law, language studies, feminism, economics and musicology.[26] Note, too,

---

[21] Not meant as a positive feature, this also was observed by Karl Rahner: "Lonergan's theological methodology seems to me to be so general that it applies equally to all sciences" (Karl Rahner, "Kritishe Bemerkungen zu B. J. F. Lonergan's Aufstaz: 'Functional Specialties in Theology,'" *Gregorianum* 51 (1970): 537-540.)

[22] Lonergan, *Method in Theology*, 364 ff. See also, Bernard Lonergan, *A Third Collection: Papers by Bernard J.F. Lonergan, S.J.*, ed. by Fred E. Crowe. Mahwah, NJ: Paulist Press, 1985.

[23] See note 19.

[24] See note 3.

[25] See, for example, note 10.

[26] Terrance J. Quinn, *The (Pre-) Dawning of Functional Specialization in Physics*. Hackensack, NJ and Singapore, World Scientific Press, in press. Sean McNelis, *Making progress in housing: a framework for collaborative research*. Oxford: Routledge, 2014.

that more than twenty years prior to Lonergan's breakthrough, Welleck and Warren's book on literature[27] practically lists the eight functional parts in the table of contents. And, philosophers of science may recall the foundational work of Arne Naess (1912-2009), father of the *Deep Ecology* movement. Naess independently identified four forward-looking groupings of tasks[28] that link closely with the four future-oriented functional specializations, namely, functional foundations, policies, systematics and communications.

E.3 Reflections: Balanced Functional Science

Besides growing in awareness of what science already has been doing, and is doing, one may wonder why a growing awareness of the eight main tasks might be important to scientific progress. That, too, is a topic in the

---

Robert Henman, "Implementing Generalized Empirical Method in Neuroscience by Functionally Ordering Tasks," *Dialogues in Philosophy, Mental and Neuroscience*, vol. 9, issue 1, June 2016. Terrance J. Quinn, "Invitation to Functional Collaboration: Dynamics of Progress in the Sciences, Technologies and Arts," *Journal of Macrodynamic Analysis*, vol. 7 (2012): 92-120. Robert Henman, "An Ethics of Philosophic Work," *Journal of Macrodynamic Analysis*, vol. 7 (2011): 44-53. Bruce Anderson, "The Nine Lives of Legal Interpretation," *Journal of Macrodynamic Analysis*, vol. 5 (2010): 30-36. John Benton, *Shaping the Future of Language Studies*. Canada: Axial Publishing, 2008. Alessandra Drage, *Thinking Woman* (Halifax: Axial Press, 2005), concluding chapters. Terrance J. Quinn, "Reflections on Progress in Mathematics," *Journal of Macrodynamic Analysis*, Vol. 3 (2003): 97-116. Philip McShane, "Elevating *Insight*, Space-Time as Paradigm Problem," *Method: Journal of Lonergan Studies*, vol. 19 (2001): 203-229. Bruce Anderson, *Discovery in Legal Decision-Making*. Dortrecht: Kluwer Academic Publishers, 1996. Philip McShane, *Lonergan's Challenge to the University and the Economy* (Lanham MD: University Press of America, 1980), chapter 5 (literary studies). Philip McShane, *Economics for Everyone - Das Jus Kapital*. Edmonton: Commonwealth Press, Edmonton, 1977 (reprinted by Axial Publishing, http://www.axialpublishing.com/). Philip McShane, *Shaping the Foundations* (Lanham, MD: University Press of America, 1976), chapter 2 on musicology - written in 1969. Philip McShane has an extensive body of work on functional collaboration, available through his website, http://www.philipmcshane.org/.

[27] René Wellek and Austin Warren, *Theory of Literature*. New York: Harcourt, Brace and World, 1942/1970.

[28] Arne Naess, "Deep Ecology and Ultimate Premises," *The Ecologist*, vol. 18, no. 4/5 (April / May 1988): 128-31.

works cited above.[29] A few paragraphs here, though, may help the reader get some first impressions; and that way also be encouraged to follow up with more detailed empirical studies. The discussion here also will bear on the question 'What is Science?' as well as on the problem of implementation.

To be able to anticipate advantages of the eightfold collaboration, we can begin by looking to two kinds of work already familiar in physics, namely, experimental physics and theoretical physics. (Eventually, these will be seen to be developmental precursors to *functional research* and *functional interpretation*). Of course, there is no rule that denies a physicist from attempting to work both as an experimental physicist and a theoretical physicist. In fact, some have done so, at least in earlier times when there was less to know and when lab technology was more elementary. Nowadays, though, experimental and theoretical physics each are highly specialized. Even being able to turn 'ON' the Large Hadron Collider is a months-long process that needs teams of technicians, each with years of appropriate education, training and skills.

Why, though, is the particle accelerator turned ON?

(W)e need collisions - and then a steady torrent of data will make its way to physicists around the world, so that the massive analysis effort can begin. [30]

What kinds of analysis? Research teams look for anomalies in data that might, for example, upset existing theories, or be clues indicating the need of revision or improved understanding of existing theories, or, may simply be puzzling. Research teams share their results and their questions with the scientific community, but especially with theoreticians. There are no strict requirements, as such, about how results need to be communicated. But, as it happens, there are well known standards and norms that have developed over the centuries and that have proven to be helpful. For instance, communications from experimental physics to the community of professional physicists usually identify at least some or all of the

---

[29] See, for example, areas mentioned in note 26.
[30] Jonathan Webb, "Large Hadron Collider restarts after two-year rebuild," in, *Science & Environment, BBC News* (April 5, 2015), http://www.bbc.com/news/science-environment-32160755.

following: Methods, Materials, Equipment; Experimental Procedure; Experimental Results; Discussion; Conclusions; References; and Appendices.

The well-known two-fold division of labor between experimental and theoretical physics has proven to be effective in various ways. Among other things, it relieves individuals from needing to simultaneously attempt to contribute to front-lines advances in both experimental work and theoretical work. Both groups, of course, are working together toward the same common goal, namely, progress in physics. Both groups necessarily have up-to-date understandings of theoretical issues, problems and questions, heuristics of GUTs and a present Standard Model. But, where experimental physics hunts for potentially significant data, theoretical physics works to explain data.

Some may object, here, to my focus here on collaboration between experimental physics and theoretical physics. For, there are ongoing debates about the role of, say, mathematics in theoretical physics. Leading proponents of string theory claim that mathematical elegance is a more or less sufficient criterion for a theory to be a correct physical theory. But, there is no need to engage in that debate here. For, whatever some might suggest about the significance of mathematical elegance, the aim of the present discussion is more elementary. I simply draw attention to what, *de facto*, is an established, familiar and regularly successful collaboration between experimental and theoretical physics. We can recall, for example, that the development of the Standard Model in the 20th century was through several decades of back-and-forth between theoreticians and experimentalists.[31] And, as is well known, that mutual exchange of results between experimental physics and theoretical physics is ongoing.

In the last two paragraphs, I was talking about physics. But, of course, a similar rapport between experiment and theory are familiar in other sciences such as chemistry, botany, and so on. Again, whatever some philosophical views might suggest, the collaboration between theoreticians in science and experimental science is a well-established

---

[31] Lochlainn O'Raifeartaigh, *The Dawning of Gauge Theory*. Princeton: Princeton University Press, 1997.

feature of scientific practice and progress. So, let's go back, now, to Lonergan's discovery of eight tasks. For, while it is well-known that experimental science and theoretical work both contribute to progress in the sciences, part of the claim about eight types of work is that: Within emerging developments in the geo-historical scientific community, experimental work and theoretical work emphasize two of eight main tasks; and all eight are mutually dependent, functionally related.

With some effort and self-attention, one finds that all eight tasks already are subtly present in the community.[32] That presence, however, is not generally adverted to. And, at this time, in individual works two or more of the tasks often are combined in *ad hoc* ways.[33]

Yet, is that kind of multi-tasking necessarily so bad? Or, is it true, perhaps, that *not* adverting to the eight distinct tasks is in some way inimical to progress? At the very least, not adverting to, and not distinguishing what in fact are eight distinct tasks means a lack of clarity about what one is doing. Mixed but non-differentiated content does not easily gain purchase in the scientific enterprise. As already discussed in several places in the book, we see this explicitly in, for example, contemporary philosophy of physics and philosophy of biology. There is high caliber scholarship in these major areas of philosophy of science. But, there are no signs, yet, of an emergent shared view or Standard Model.[34]

If more or less free multi-tasking is inimical to progress, how might adverting to, and growing in a working knowledge of the eightfold groupings of tasks be expected to favor progress? Collaboration between experimental and theoretical work has been conspicuously generative of cumulative and progressive results in the sciences. In the basic and applied

---

[32] See, for example, work in note 26.

[33] As will emerge, "the distinction and division are needed to curb totalitarian ambition" (Bernard Lonergan, *Method in Theology* (London: Darton, Longman and Todd, 1972/73/75), 137).

[34] Robert Batterman, ed., *The Oxford Handbook of the Philosophy of Physics*. Oxford: Oxford University Press, 2013; Dean Rickles, ed., *The Ashgate Companion to Contemporary Philosophy of Physics*. Burlington, VT: Ashgate Publishing Company, 2008; and John D. Barrow, Paul C. W. Davies, Charles L. Harper, Jr., *Science and Ultimate Reality, Quantum Theory, Cosmology, and Complexity*. Cambridge: Cambridge University Press, 2004.

sciences and philosophy of science, might not being in control of all eight tasks be expected to ground a considerable increase in both economy and effectiveness in overall collaborations?

Further advantages can be anticipated by observing that developments in the scientific community have tended to be within tunnelings of discipline-specific work. This is not to suggest that developments in chemistry, or psychology, or mathematics, or philosophy, or other differentiations have not been part of major developments and progress, or that differentiated expertise is not normative. But, it is also true that at this time, generally, there is a lack of consensus about how best to share results with, or take advantage of developments in other tunnelings. And,

the discussion of interdisciplinarity in the philosophy of science is still in its infancy.[35]

Still, if the hypothesis that there are eight functionally related tasks "holds up under philosophic scrutiny,"[36] as I suggest it will, an eightfold functional collaboration can be expected to provide a controlling context, within which the present meaning of "interdisciplinarity"[37] can be expected to shift, or even become obsolete within an omni-disciplinary heuristics of emerging functional collaboration.

Eventually, a growing working knowledge of the eight tasks can be expected to go well beyond any present or near-future descriptions. We can begin to envision a future, then, in which the eight tasks have become luminously and explanatorily differentiated, at the level of the times. In that future stage of meaning,[38] there will be not merely eight types of work obscurely present in the academic enterprise. What will emerge, instead, will be eight explicit *functional specialties* in a global functional collaboration. This will be an extraordinary shift in control of meaning in

---

[35] Till Grüne-Yanoff, "Introduction: Interdisciplinary model exchanges," (2014): 59.

[36] Douglas, 55.

[37] See note 35.

[38] There will be a "third stage of meaning" (Lonergan, *Method in Theology*, sec. 3.10, 85-99.

the academic community.[39] Eight progress-oriented functional specialties will be

distinct and separate stages in a single process from data to ultimate results.[40]

And so we come again to the question, What is Science? In mature science, eight tasks will be luminously differentiated within a control of meaning attained through generalized empirical method. Mature science, then, will involve conception, affirmation and implementation[41] of a normative omni-disciplinary eightfold division of labor.

Functional specialization will be its own differentiation. Some evidence for this is indirect. Lonergan was a genius, and it is a matter of record that he discovered functional specialization later in life, decades after already having attained remarkable and accelerating control of meaning in the sciences, philosophy and theology. For direct evidence, we can appeal to personal experience. Again, there is the need of self-attention, within established zones of expertise.[42] Without such efforts, talk of functional specialization, even if interesting, will remain story-like, with little chance of contributing to progress in the modern sciences and philosophy of science.

However, present disorientations in the Academy make empirical probabilities of even descriptive self-attention approximately zero. So, what might we do to promote the emergence of either generalized empirical method, or functional specialization, or both? Surprisingly,

---

[39] Present-day confusions will "recede in the measure that linguistic *feedback is achieved, that is, in the measure that* explanations and statements provide the sensible presentations for the insights that effect further developments of thought and language" (Lonergan, *Method in Theology*, lines 12-15, 92). The text in italics is missing from editions so far, including Toronto: University of Toronto Press, 1990/94/96, 2003. It seems that the missing text accidentally got dropped from the printer's copy of the final MS. See archival document 52300DTE060/69-5A, page 3, available at: http://www.bernardlonergan.com/archive.php. Thanks to Philip McShane and Pat Brown for drawing this to my attention and for the link to the archive file.

[40] Lonergan, *Method in Theology*, 136.

[41] See note 5.

[42] Again, see note 26, for various preliminary efforts along these lines.

perhaps, it would seem that promoting functional specialization will be a strategic way to go. Even though a lack of control of meaning is prevalent in the present-day academy, and even if, for now, adverting to operations is a low probability event, the eight main divisions of labor already are being gradually forced into view by the pressures of history.

Before too long there will be efforts toward adjusting patterns of scholarship within a normative eightfold collaboration. No doubt, early efforts will be messy and not well differentiated. The effectiveness of the eightfold division of labor, however, soon will be undeniable and welcome. By that time, even if there are still hold-outs, it will be evident to the main Academic community that not collaborating within the normative pattern would be analogous to attempting to do chemistry while ignoring the discovery of the periodic table.

What, though, of generalized empirical method? As history reveals, emergence of the balanced empirical method may not happen any time soon. It is, in a sense, too much to expect, well, at least not without some kind of auxiliary support.[43] But, within the supporting context of even preliminary efforts toward approximating an eightfold division of labor, the need for control of meaning, self-attention, and the need to determine metaphysical equivalences soon would be brought out into the open. Following on those same first approximations to functional collaboration, we can expect jumps in probabilities of emergence of increasingly efficient eightfold schemes of recurrence. In turn, and in turning,[44] growth in eightfold cyclic collaboration in the academy will bring forth still further developments, differentiations and control of meaning. That way, empirical method in the academy gradually will mature toward being *functionally differentiated generalized empirical method.*

---

[43] Functional specialization will be "a specialized auxiliary ever ready to offset every inter-ference with either intellect's unrestricted finality or with its essential detachment and disinterestedness" (*CWL3*, 747).

[44] See Fig. 1, W5, in Philip McShane, "The Importance of Rescuing *Insight*," in John J. Liptay Jr. and David S. Liptay, *The Importance of Insight* (Toronto: University of Toronto Press, 2007), 204.

Index

Printed in the United States
By Bookmasters